NEUROMETHODS

Series Editor
Wolfgang Walz
University of Saskatchewan
Saskatoon, SK, Canada

For further volumes:
http://www.springer.com/series/7657

Neuromethods publishes cutting-edge methods and protocols in all areas of neuroscience as well as translational neurological and mental research. Each volume in the series offers tested laboratory protocols, step-by-step methods for reproducible lab experiments and addresses methodological controversies and pitfalls in order to aid neuroscientists in experimentation. *Neuromethods* focuses on traditional and emerging topics with wide-ranging implications to brain function, such as electrophysiology, neuroimaging, behavioral analysis, genomics, neurodegeneration, translational research and clinical trials. *Neuromethods* provides investigators and trainees with highly useful compendiums of key strategies and approaches for successful research in animal and human brain function including translational "bench to bedside" approaches to mental and neurological diseases.

Neurosurgical Robotics

Edited by

Hani J. Marcus

The National Hospital for Neurology and Neurosurgery, Queen Square, London, UK

Christopher J. Payne

John A. Paulson School of Engineering and Applied Sciences, Harvard University, Cambridge, MA, USA

Humana Press

Editors
Hani J. Marcus
The National Hospital for Neurology
and Neurosurgery, Queen Square
London, UK

Christopher J. Payne
John A. Paulson School of Engineering
and Applied Sciences
Harvard University
Cambridge, MA, USA

ISSN 0893-2336 ISSN 1940-6045 (electronic)
Neuromethods
ISBN 978-1-0716-0995-8 ISBN 978-1-0716-0993-4 (eBook)
https://doi.org/10.1007/978-1-0716-0993-4

Series Preface

Experimental life sciences have two basic foundations: concepts and tools. The *Neuromethods* series focuses on the tools and techniques unique to the investigation of the nervous system and excitable cells. It will not, however, shortchange the concept side of things as care has been taken to integrate these tools within the context of the concepts and questions under investigation. In this way, the series is unique in that it not only collects protocols but also includes theoretical background information and critiques which led to the methods and their development. Thus it gives the reader a better understanding of the origin of the techniques and their potential future development. The *Neuromethods* publishing program strikes a balance between recent and exciting developments like those concerning new animal models of disease, imaging, in vivo methods, and more established techniques, including, for example, immunocytochemistry and electrophysiological technologies. New trainees in neurosciences still need a sound footing in these older methods in order to apply a critical approach to their results.

Under the guidance of its founders, Alan Boulton and Glen Baker, the *Neuromethods* series has been a success since its first volume published through Humana Press in 1985. The series continues to flourish through many changes over the years. It is now published under the umbrella of Springer Protocols. While methods involving brain research have changed a lot since the series started, the publishing environment and technology have changed even more radically. Neuromethods has the distinct layout and style of the Springer Protocols program, designed specifically for readability and ease of reference in a laboratory setting.

The careful application of methods is potentially the most important step in the process of scientific inquiry. In the past, new methodologies led the way in developing new disciplines in the biological and medical sciences. For example, Physiology emerged out of Anatomy in the nineteenth century by harnessing new methods based on the newly discovered phenomenon of electricity. Nowadays, the relationships between disciplines and methods are more complex. Methods are now widely shared between disciplines and research areas. New developments in electronic publishing make it possible for scientists that encounter new methods to quickly find sources of information electronically. The design of individual volumes and chapters in this series takes this new access technology into account. Springer Protocols makes it possible to download single protocols separately. In addition, Springer makes its print-on-demand technology available globally. A print copy can therefore be acquired quickly and for a competitive price anywhere in the world.

Saskatoon, SK, Canada   *Wolfgang Walz*

Preface

Neurosurgery is arguably the most technology-intensive surgical discipline. New technologies have preceded many of the major advances in operative neurosurgical techniques, including microsurgery, endoscopy, and image guidance. Robotics has the potential to transform neurosurgery by enabling more precise and delicate interventions. The first report of robot-assisted neurosurgery was in 1985, when a modified industrial robot was used to define the trajectory of a brain biopsy. Since this early description, numerous other neurosurgical robots have been developed. More recently, neurosurgical robots have begun to be introduced into mainstream clinical practice—everything from deep brain stimulation to pedicle screw placement.

Neurosurgical Robotics is the first book dedicated to providing an introduction to engineers and healthcare professionals working in this dynamic field. The book is divided into two parts: Part I introduces the basic engineering concepts that underpin surgical robotics, the various types of robotic platforms, and the process by which such systems make their way from bench to bedside. Part II explores popular applications of surgical robots in neurosurgery within each subspecialty and concludes with a vision for future direction in the development of neurosurgical robotic systems.

An important consideration with neurosurgical robotics is that surgical innovation and evaluation should occur together in a logical and ordered manner. Where possible, we have asked contributors to discuss the evidence for robot-assisted procedures and to highlight where this is currently lacking. We would encourage all readers involved in this rapidly evolving field to actively participate in providing such evidence; it is only by doing so that we can collectively ensure neurosurgical robotic systems are contributing to safe, effective, and efficient surgery.

London, UK   *Hani J. Marcus*
Cambridge, MA, USA   *Christopher J. Payne*

Contents

Contributors

AMIR BAGHDADI • *Project neuroArm, Department of Clinical Neurosciences and Hotchkiss Brain Institute, University of Calgary, Calgary, AB, Canada*

DANIEL BAUTISTA-SALINAS • *Hamlyn Centre for Robotic Surgery, Imperial College London, London, UK*

TREVOR L. BRUNS • *Vanderbilt University, Nashville, TN, USA*

ASWIN CHARI • *Great Ormond Street Hospital for Children NHS Foundation Trust, London, UK; University College London Great Ormond Street Institute of Child Health, London, UK*

DORIAN CHAUVET • *Neurosurgery, Rothschild Foundation, Paris, France*

WILLIAM T. COULDWELL • *Department of Neurosurgery, Clinical Neurosciences Center, University of Utah, Salt Lake City, UT, USA*

EMMANOUIL DIMITRAKAKIS • *Wellcome/EPSRC Centre for Interventional and Surgical Sciences, University College London, London, UK*

GEORGE DWYER • *Wellcome/EPSRC Centre for Interventional and Surgical Sciences, University College London, London, UK*

STEFANO GALVAN • *Department of Mechanical Engineering, Imperial College London, London, UK*

STEVEN S. GILL • *North Bristol NHS Trust, Bristol, UK*

DAVID D. GONDA • *Rady Children's Hospital of San Diego, San Diego, CA, USA; University of California, San Diego, CA, USA*

STÉPHANE HANS • *Head and Neck Department, Hopital Européen Georges Pompidou, Paris, France*

NADER M. HEBELA • *Department of Neurosurgery, Cleveland Clinic, Abu Dhabi, United Arab Emirates*

HAMIDREZA HOSHYARMANESH • *Project neuroArm, Department of Clinical Neurosciences and Hotchkiss Brain Institute, University of Calgary, Calgary, AB, Canada*

JOHN PAUL G. KOLCUN • *Department of Neurological Surgery, Miller School of Medicine, University of Miami, Miami, FL, USA*

BORNALI KUNDU • *Department of Neurosurgery, Clinical Neurosciences Center, University of Utah, Salt Lake City, UT, USA*

SANJU LAMA • *Project neuroArm, Department of Clinical Neurosciences and Hotchkiss Brain Institute, University of Calgary, Calgary, AB, Canada*

JOHN Y. K. LEE • *Department of Neurosurgery, Hospital of the University of Pennsylvania, Philadelphia, PA, USA*

MICHEL LEFRANC • *GRECO: Groupement de Recherches et d'Etudes en Chirurgie Robotisée, Neurosurgery Department, CHU Amiens Picardie, Jules Verne University of Picardie, Amiens, France*

HANI J. MARCUS • *The National Hospital for Neurology and Neurosurgery, Queen Square, London, UK*

ELOISE MATHESON • *Department of Mechanical Engineering, Imperial College London, London, UK*

CATHERINE MORAN • *North Bristol NHS Trust, Bristol, UK*

DAO M. NGUYEN • *Section of Thoracic Surgery, Division of Cardiothoracic Surgery, Department of Surgery, University of Miami, Miami, FL, USA; UHealth Tower, Miami, FL, USA*

CHRISTOPHER J. PAYNE • *John A. Paulson School of Engineering and Applied Sciences, Harvard University, Cambridge, MA, USA*

MARLENE PINZI • *Department of Mechanical Engineering, Imperial College London, London, UK*

JILLIAN PLONSKER • *Rady Children's Hospital of San Diego, San Diego, CA, USA; University of California, San Diego, CA, USA*

SYED S. RAZI • *Section of Thoracic Surgery, Division of Cardiothoracic Surgery, Department of Surgery, University of Miami, Miami, FL, USA*

ANDRIA A. REMIREZ • *Vanderbilt University, Nashville, TN, USA*

FERDINANDO RODRIGUEZ Y BAENA • *Department of Mechanical Engineering, Imperial College London, London, UK*

FLORIAN ROSER • *Department of Neurosurgery, Cleveland Clinic, Abu Dhabi, United Arab Emirates*

MARGARET F. ROX • *Vanderbilt University, Nashville, TN, USA*

PAUL T. RUSSELL • *Vanderbilt University, Nashville, TN, USA*

RICCARDO SECOLI • *Department of Mechanical Engineering, Imperial College London, London, UK*

TAKU SUGIYAMA • *Project neuroArm, Department of Clinical Neurosciences and Hotchkiss Brain Institute, University of Calgary, Calgary, AB, Canada*

PATRICIA ZADNIK SULLIVAN • *Department of Neurosurgery, Hospital of the University of Pennsylvania, Philadelphia, PA, USA*

GARNETTE R. SUTHERLAND • *Project neuroArm, Department of Clinical Neurosciences and Hotchkiss Brain Institute, University of Calgary, Calgary, AB, Canada*

M. ZUBAIR TAHIR • *Great Ormond Street Hospital for Children NHS Foundation Trust, London, UK; University College London Great Ormond Street Institute of Child Health, London, UK*

MARTIN M. TISDALL • *Great Ormond Street Hospital for Children NHS Foundation Trust, London, UK; University College London Great Ormond Street Institute of Child Health, London, UK*

NESTOR VILLAMIZAR • *Section of Thoracic Surgery, Division of Cardiothoracic Surgery, Department of Surgery, University of Miami, Miami, FL, USA*

VANI VIRDYAWAN • *Department of Mechanical Engineering, Imperial College London, London, UK*

KHUSHI VYAS • *Hamlyn Centre for Robotic Surgery, Imperial College London, London, UK*

MICHAEL WANG • *Department of Neurological Surgery, Miller School of Medicine, University of Miami, Miami, FL, USA*

THOMAS WATTS • *Department of Mechanical Engineering, Imperial College London, London, UK*

ROBERT J. WEBSTER III • *Vanderbilt University, Nashville, TN, USA*

WILLIAM C. WELCH • *Department of Neurosurgery, Hospital of the University of Pennsylvania, Philadelphia, PA, USA*

GUANG-ZHONG YANG • *Institute of Medical Robotics, Shanghai Jiao Tong University, Shanghai, China*

DANDAN ZHANG • *Hamlyn Centre for Robotic Surgery, Imperial College London, London, UK*

Part I

Development

Chapter 1

Basic Concepts in Robotics

Christopher J. Payne, George Dwyer, Emmanouil Dimitrakakis, and Hani J. Marcus

Abstract

This chapter aims to provide a practical introduction to the basic concepts in robotics relevant to neurosurgical robotic systems. We start by introducing different robotic architectures and discussing how these contrasting fundamental approaches are suited to different neurosurgical applications. We then examine the different actuators and mechanisms that give robots motion. Next, we look at the different sensing technologies that allow robots to sense themselves and their external environment. With some knowledge of robot architectures, actuation, and sensing, we then examine how these elements can be combined together into a controllable robotic system. Finally, we introduce the different user-interface paradigms applicable to neurosurgical robots.

Key words Robotics, Actuators, Sensors, Control, Kinematics, Dynamics, Mechanisms, Engineering

1 Introduction

This chapter is intended for both clinicians and engineers wanting to garner a basic understanding of the fundamental concepts behind the design and operation of medical robotic systems. The broad field of robotics is colossal, incorporating many disciplines of engineering including; mechanical engineering, electrical engineering, and computer science. The reader will appreciate that it would be impractical to review such a huge field in significant depth within this book. Thus, the scope of this chapter is to provide a brief, high-level introduction to some of the very core concepts within robotics that are of significance to neurosurgical robotic systems. There are many excellent textbooks and journal articles that cover the broader field of robotics which are referenced throughout this chapter for readers to consult should they wish to develop a more in-depth understanding of specific technical aspects. In particular, the *Springer Handbook of Robotics* [1] comprehensively covers all facets of modern robotics. In addition to introducing basic robotics concepts, this chapter also serves to introduce some of the key

Hani J. Marcus and Christopher J. Payne (eds.), *Neurosurgical Robotics*, Neuromethods, vol. 162,
https://doi.org/10.1007/978-1-0716-0993-4_1, © Springer Science+Business Media, LLC, part of Springer Nature 2021

terminology used by roboticists and engineers. Key terms in robotics are highlighted throughout, and a corresponding glossary is provided at the end of this chapter.

It is often the case that concepts in robotics can be discussed in abstract and general terms. Since this book is specifically focused on neurosurgical robotics and intended for a broad audience of both engineers and clinicians, this chapter aims to present basic robotics concepts in a concrete and practical manner. One way in which we aim to achieve this is by introducing theoretical concepts through exemplar surgical robotic systems. This approach also gives us the opportunity to analyze neurosurgical robotic systems from a design perspective. As we step through the basic concepts in this way, we can relate these conceptual ideas to how robots are designed to meet the functional requirements of real-world neurosurgical applications.

2   What is a Robot?

In the broadest terms, robots are programmable machines capable of performing autonomous tasks. At the highest level, all robots will incorporate some form of actuator, sensor and control system. Like biological muscles, **actuators** are the devices and subsystems within a robot that generate forces and motion. **Sensors** enable robotic systems to sense themselves and/or their surrounding environment, much like our own sensory nervous system. **Controllers** are analogous to a brain; they acquire and process sensory input information and output commands to the actuation system. These individual components are combined together and organized into an integrated robot.

3   Types of Robot

Robotic systems are commonly classified according to their underlying physical architecture. In this chapter we shall discuss two broad types of robotic system: **rigid robotic systems** and **continuum robots**.

3.1   Rigid Robotic Systems

The common robot architectures utilized in medical applications tend to be based on conventional robotic manipulator designs that bear some resemblance to a human arm. Like a human arm, robot arms incorporate a series of **links** that are connected by **joints** which can rotate (**revolute joints**) or in some cases translate (**prismatic joints**). This chain of links and joints is typically referenced to a **base** to which the robot is anchored. Coordinated rotation and/or translation of these individual joints allows the robot arm to make articulated movements relative to the base. At the most

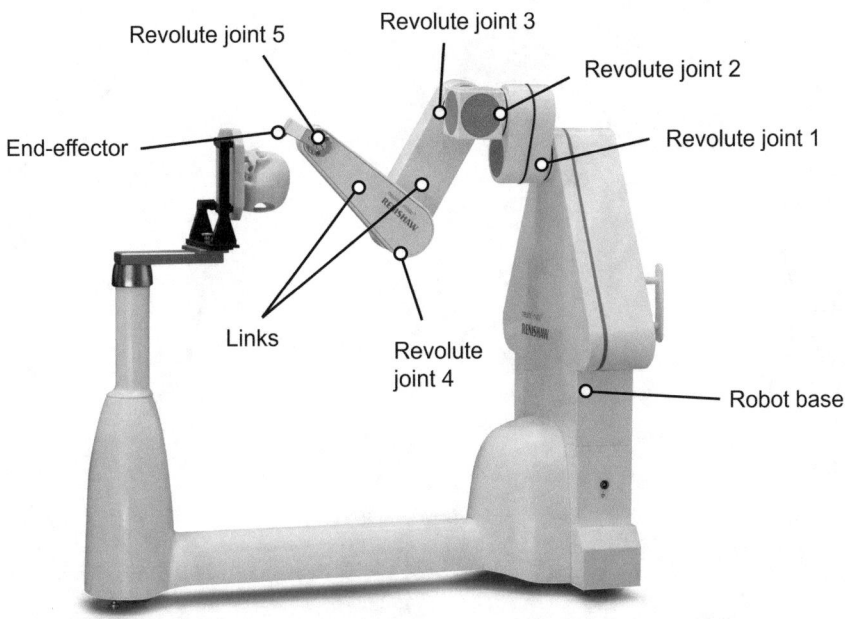

Fig. 1 Basic elements of a rigid robotic system as illustrated on the Renishaw Neuromate® with permission, © Renishaw plc

distal region of a robot arm is the **end-effector**, in an industrial robot this might be a robotic hand or welding instrument, in a medical robot the end-effector might be an interventional tool or imaging device. Figure 1 illustrates these basic elements of a robotic manipulator. A further property of interest for robotic manipulators is the number of **degrees-of-freedom (DoF)** they possess. In robotics, a degree-of-freedom refers to an independent mode by which a robot can move. The number of DoFs that a robot has will determine to what extent the robot arm can position itself about the surrounding space. The **reachable workspace** of a robot refers to a volume that describes the complete physical region that the robot end-effector can reach (*see* Fig. 2). Of greater interest is the **dexterous workspace**, which describes the physical volume in which the robot can move with each DoF, so that the end-effector can be arbitarily positioned and oriented.

The rigid robotic systems discussed thus far can be subcategorized into serial and parallel architectures. **Serial robots** are designed with each link and joint organized sequentially, much like a human arm. These robots can provide a large dexterous workspace allowing the robot to reach over large distances and a wide range of orientations. Conversely, **parallel robots** have multiple chains of links and joints that are coupled together to support a single end-effector.

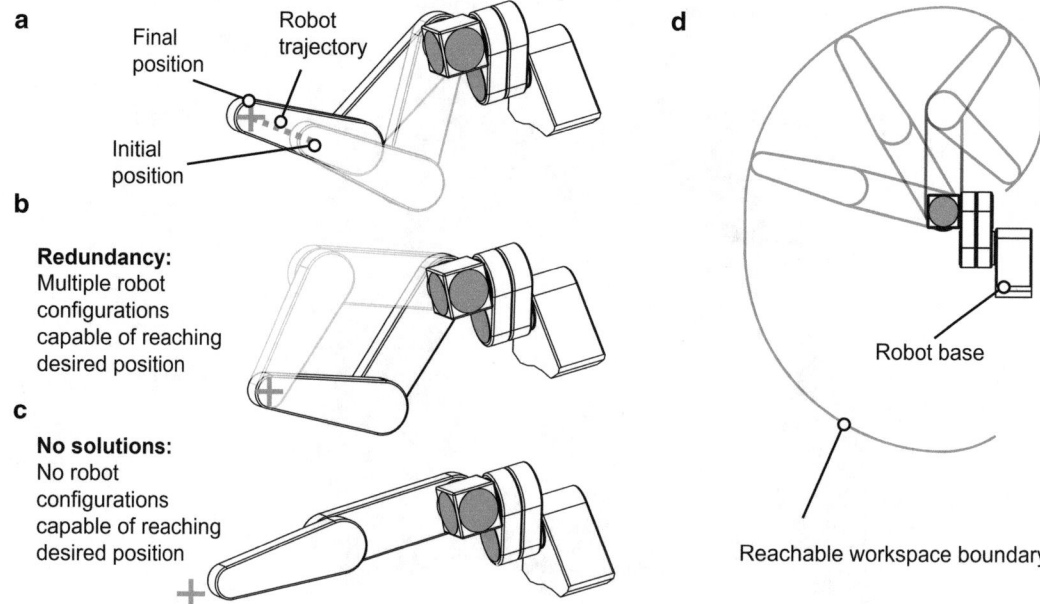

Fig. 2 (**a**) Illustration of a robot moving through a trajectory from an initial position to a new position. (**b**) Illustration of redundancy in a robotic system in which multiple kinematic configurations of the robotic system can reach the same point. (**c**) Example of when no kinematic configurations exist for a robot commanded to reach a position beyond its workspace. (**d**) Illustrative example of the outer boundary limit of the reachable workspace for a robotic manipulator

Parallel architectures have some advantages over serial robotic systems; they can exhibit superior positioning accuracy and accelerate quicker. Although robots can move very precisely, no robotic joint is perfect and will exhibit some degree of positioning error. In a serial architecture, inaccuracies in the position of each joint will accumulate along the chain of links and joints leading to a substantial error at the end-effector position. Parallel robots, on the other hand, can be designed to have fewer links in each chain and the errors that do exist within a chain are averaged across the robot. The parallel linkage architecture of these robots is also innately stiffer when compared to a single linkage chain, and this improved mechanical stability also serves to reduce positioning errors. A drawback of parallel robots is that their workspace is inherently constrained by their geometrical configuration, since the individual linkage chains in a parallel robot can potentially collide with each other.

The choice of robot structure is often determined by the target application. To take two examples, the NeuroArm robotic system (University of Calgary [2]) incorporates serial manipulators whereas the SpineAssist® device (Medtronic plc, Minneapolis, USA) is based on a parallel architecture (Fig. 3). The NeuroArm

a

Serial robot: NeuroArm

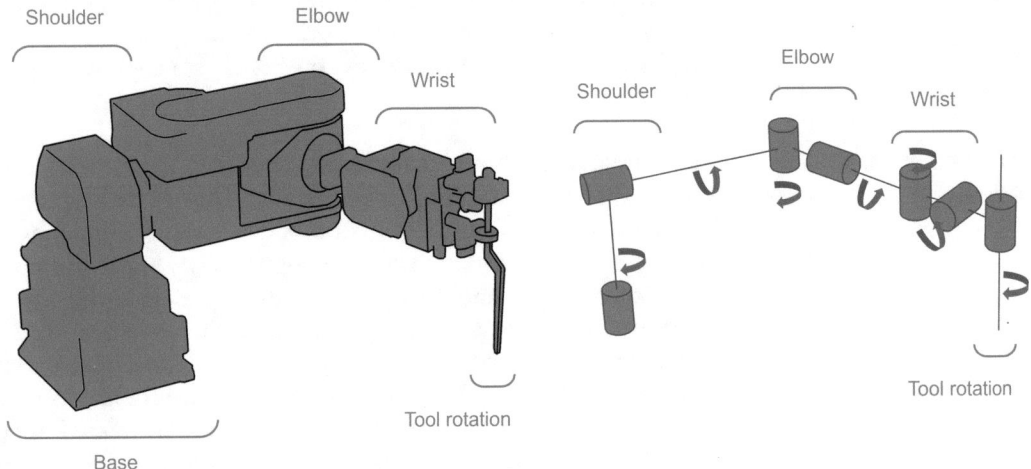

b

Parallel robot: SpineAssist

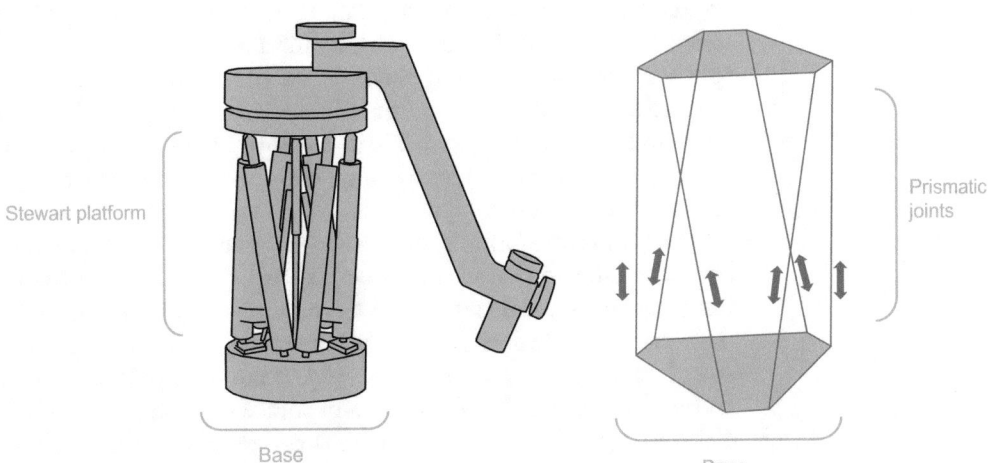

Fig. 3 (**a**) The NeuroArm system [2] as an example of a serial robot architecture and (**b**) the SpineAssist®
device (Medtronic plc, Minneapolis, USA) as an example of a parallel robot architecture

system is designed to directly operate on a patient in the same way
that a surgeon would do manually. The surgeon controls a user
interface which reads in their input motions and recreates them on
the robotic manipulator. In this application, an anthropomorphic
robotic arm based on a serial architecture is a preferable choice
because its underlying structure mirrors that of the human arm it
intends to emulate. Furthermore, this serial robot chain can

provide a large workspace at the end-effector location, allowing for dexterous articulation of the interventional tool at the surgical site. The SpineAssist® robot is designed for image-guided needle placement where positional accuracy is a primary design input requirement; thus, a parallel robot architecture is an appropriate design choice.

3.2 Continuum Robots

We have discussed robotic manipulator arms that are articulated in nature and rely on rotations and/or translations of rigid links in order to achieve motion. Rigid robotic manipulators have been extensively adopted in manufacturing applications, in part because their inherent rigidity ensures they can achieve precise and controllable positioning. This same virtue has great utility in neurosurgical applications; indeed, both the NeuroArm and the SpineAssist® systems use rigid manipulators to ensure good positional accuracy. However, rigid robotic architectures also incorporate some limitations that can be problematic in surgical environments. Firstly, their rigid nature makes them potentially hazardous for directly interacting with patients since a collision could cause significant injury. Secondly, while rigid robots are excellent at positioning instrumentation outside of the body, it is technically challenging and often impractical to miniaturize articulated robot designs to the physical scales that can enable endoscopic interventions. An alternative approach to rigid robotic systems are continuum robot designs that achieve dexterity through physical deformation [3]. These tentacle-like devices can achieve smooth, continuous motions, making them ideal for navigating confined spaces and tortuous paths within the body.

Continuum robotic systems can be catheter-like, in which tendons are used to deflect a flexible tool tip. Highly dexterous continuum robots can be fabricated by incorporating multiple tendons that can each be independently-controlled to provide complex, coordinated snake-like motions [4]. Like conventional flexible endoscopes, these designs can incorporate cameras, light sources, and channels for interventional instrumentation. Other designs utilize actuatable flexible backbones to create snake-like manipulators [5]. Similar feats of dexterous manipulation can be achieved using concentric tube robots [6, 7]. These systems are comprised of nested tubes that each have some degree of preformed curvature and are fabricated from a highly elastic material such as Nitinol. A robotic control system can independently translate and rotate the individual nested tubes which causes coordinated deflection and snaking motions of the overall tubular assembly. Since these robots are tubular, interventional instrumentation can be passed through the entire tube assembly to enable complex robotic systems. Needle steering robotic systems are another category of continuum robot relevant to neurosurgery. Most interventional needles are relatively stiff and designed to remain straight when inserted into soft tissue.

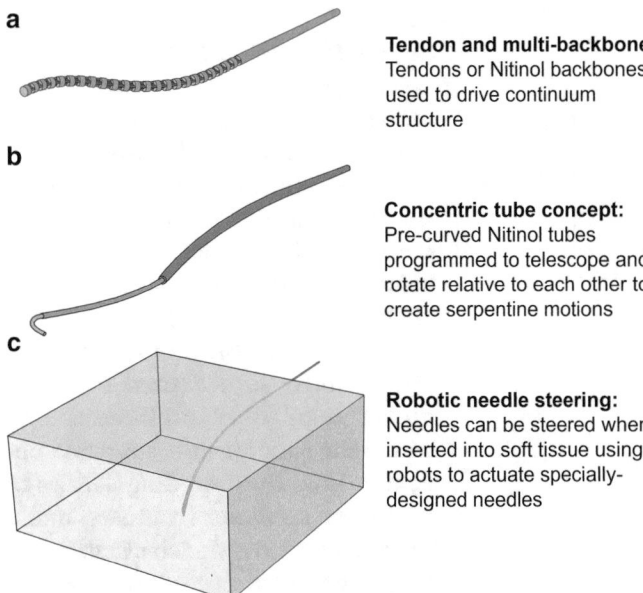

Tendon and multi-backbone: Tendons or Nitinol backbones used to drive continuum structure

Concentric tube concept: Pre-curved Nitinol tubes programmed to telescope and rotate relative to each other to create serpentine motions

Robotic needle steering: Needles can be steered when inserted into soft tissue using robots to actuate specially-designed needles

Fig. 4 Examples of continuum-based robotic architectures. (**a**) Tendon and multi-backbone-based systems, (**b**) concentric tube robots, and (**c**) robotic needle steering systems

Robotics researchers have developed flexible needles designed to deflect in a predictable manner when inserted into soft tissue so that the needle insertion trajectory can be preplanned [8–10]. In neurosurgery, such an approach can allow for planned needle insertion into a specific region of the brain while avoiding critical structures. These continuum robot concepts are illustrated in Fig. 4.

4   Kinematics and Dynamics

Kinematics is a general field of mechanics concerned with the motion of bodies without consideration of the forces that cause this motion. In surgical robot design, it is important to analyze how a robotic manipulator travels through space when it is commanded to move to a new position. While robot kinematic analysis methods are well established for rigid robotic systems, kinematics calculations are much more challenging to perform for continuum-based robot architectures. In our discussion on kinematics and dynamics, we will solely focus on rigid robotic architectures.

In studying the kinematics of an articulated robotic manipulator, we want to relate how the position of each robot joint (whether a rotating joint or a translating joint) corresponds to the position and orientation of the robot end-effector. If we know the orientation of each joint in our robotic manipulator, we can compute the end position and orientation of the robot end-effector using geometric relations. Roboticists refer to this procedure as calculating the **forward kinematics** of a robot. If we wish to perform the reverse process, that is, to determine the required positions of each individual joint for a given end position of the robot, we must compute the **inverse kinematics**.

Calculating the inverse kinematics of a robot is considerably more complicated than computing the forward kinematics. In some cases, a desired robot pose is impossible to achieve (there are no solutions) and in other instances, multiple or even infinite solutions are possible (this is known as **kinematic redundancy**). The former condition may also be the result of **singularities**, where a robot cannot move in a desired direction regardless of how its joints move. Singularities can be seen at the boundary of the workspace where the robot is fully stretched out and it is unable to move any further from the base. Singularities can also arise when two or more joints line up along their axes, and careful programming must be carried out to avoid such events occurring. Robots typically need to be programmed according to the inverse kinematic calculations. For example, in the case of a robotic needle guidance system, the needle insertion trajectory will be predetermined by preoperative imaging data, and the robot joint positions must subsequently be calculated based on this desired trajectory.

Kinematics calculations can be used to plan the trajectory through which a robot will move, however kinematic calculations alone do not allow the velocity or force of the robot to be controlled precisely. **Dynamics** is a field of mechanics that relates how the forces acting on a body relate to its motion. In robotics, **forward dynamics** calculations analyze how a robot will move for a given force or torque input at each joint. Conversely, computation of the **inverse dynamics** will determine the torques and/or forces required to achieve a desired acceleration profile. To make dynamics calculations, the robot designer must understand not only the robot structure but also the mass and inertial properties of each joint within the robotic system.

If readers wish to garner a more comprehensive understanding of robot kinematics and dynamics, we recommend the following textbooks:

- Waldron K, Schmiedeler J (2016) Kinematics. In: Siciliano B, Khatib O (eds) Springer handbook of robotics. Springer, New York, pp 9–33

- Featherstone R, Orin DE (2016) Dynamics. In: Siciliano B, Khatib O (eds) Springer handbook of robotics. Springer, New York, pp 35–65

- Craig JJ (2005) Introduction to robotics: mechanics and control, 3rd edn. Pearson, Upper Saddle River, NJ

- Spong MW, Vidyasagar M (2008) Robot dynamics and control. John Wiley & Sons, Hoboken, NJ

5 Actuators and Mechanisms

We have reviewed the very basics of robotic systems and contrasted different types of robotic architecture. We have also considered how robotic manipulators can be programmed to move through space, but without any consideration for how this motion is generated. Actuators are components within a robot that generate motion. Robotic manipulators such as those we have examined in the previous section will warrant multiple actuators that independently create motion at each joint within the robot. Actuators typically do not directly move a robotic joint, but instead motion must be transmitted through a mechanism. Mechanisms transform one form of input motion or force into an output motion or force. For example, it is often the case that an actuator cannot generate sufficient torque to rotate a robotic joint, so a mechanism (such as a gearbox) is used to amplify the torque delivered to the joint. In other cases, it might be necessary to transform the type of motion. For example, we may wish to use an electric motor to actuate a prismatic joint and this would warrant a mechanism that converts rotary motion to linear motion. There is an abundance of actuation technologies and mechanisms that have been designed for robotic applications. In this section we shall review some key exemplar technologies relevant to neurosurgical robotic systems.

5.1 Electromagnetic-Based Motors

The most common actuators used in robotics are **electromagnetic motors**. At the most fundamental level, electromagnetic motors are machines that convert electrical energy into mechanical energy using magnetic fields. Most electromagnetic motors in robotics applications are **direct current** (DC), meaning that the electrical input to the motor flows in a single direction. The most common DC electromagnetic motors are rotary-based and incorporate two key subassemblies: a **stator** (static components) and a **rotor** (rotating components). Such rotary motors can be stratified into **brushed** and **brushless** designs.

In a brushed DC electromagnetic motor, the stator incorporates magnets (typically permanent magnets) mounted in an outer housing (*see* Fig. 5). A rotor assembly is mounted within this outer housing and incorporates the motor **armature**. The armature is an electromagnet—a series of electrical windings around an iron core that can generate an electromagnetic field when supplied with an electrical current. Brushed DC motors also incorporate a **commutator,** an electrical switch that reverses the polarity of the electrical current being supplied to the armature as it rotates. Since the armature is rotating, electrical power is conducted to the armature using sliding contacts known as brushes. When the motor armature is energized it generates an opposing electromagnetic field to the magnetic field generated by the stator magnets. This

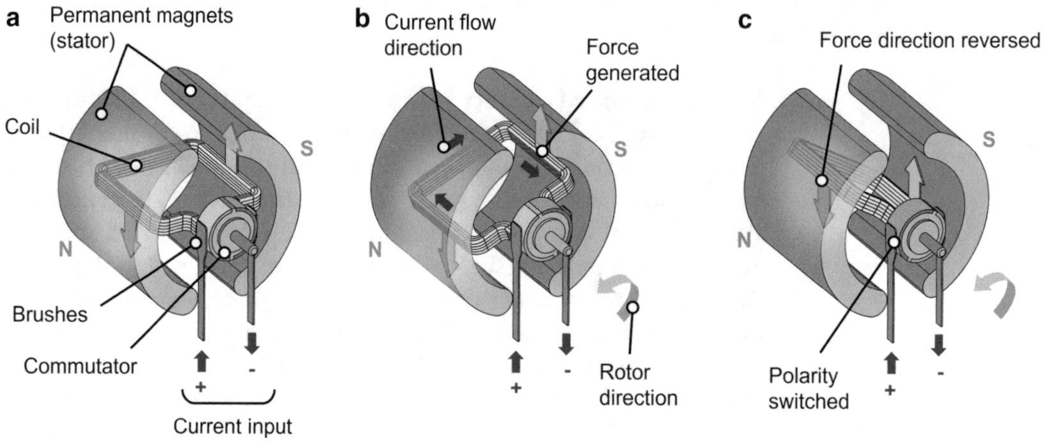

Fig. 5 Phases of operation in a brushed DC electric motor: (**a**) motor coil is energized generating a torque, (**b**) rotor rotates anticlockwise and then (**c**) commutator switches the current polarity in the coil, which reverses the coil magnetic field direction and perpetuates rotation of the rotor

electromagnetic field induces a repulsion and attraction force on opposing sides of the rotor. These opposing forces induce a torque on the rotor which causes it to then rotate. Since the stator assembly has opposing magnetic poles, the commutator switches the polarity of the electrical current that is supplied to the armature every half revolution of the motor. Thus, the electromagnetic field generated by the armature always opposes the magnetic field of the stator assembly. This commutation process ensures a unidirectional torque is induced in the rotor, so long as a direct current is supplied to the motor. The motor direction can then be reversed by switching the polarity of the electrical current supply. The torque generated by the motor is proportional to the current supplied and the motor speed can be controlled on this basis. The main advantage of these motors is that they can be driven with a simple DC input and controlled with relative ease. However, the brush-based commutator design adds friction which reduces the motor efficiency and life span due to wear.

Brushless DC electromagnetic motor designs achieve commutation of the armature using electrical switching rather than a mechanical commutator which obviates the need for contact brushes. Brushless motors are thus configured in the opposite way to brushed motors: the coil-based armature is the stator and the rotor incorporates a permanent magnet assembly (*see* Fig. 6). The stator armature coils are selectively energized in a sequence to attract the permanent magnet in the rotor assembly and generate rotary motion. Brushless motors cannot be rotated with a constant current, they require a controller to manage sequential charging of the armature coils. Brushless motors must incorporate sensors to detect the angular position of the rotor to correctly synchronize energizing of the coils. Since no brush-based sliding contacts are

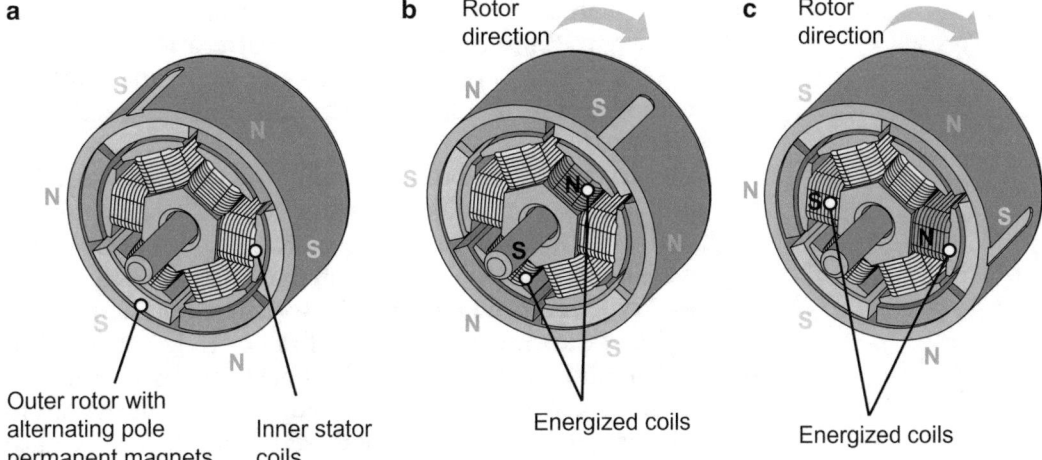

Outer rotor with
alternating pole
permanent magnets

Inner stator
coils

Energized coils

Energized coils

Fig. 6 Phases of motion in a brushless DC electromagnetic motor: (**a**) static motor, (**b**) opposing stator coils energized to induce a torque on the rotor, and (**c**) adjacent coil pairing energized to perpetuate rotation of the rotor

required in a brushless motor, they have greater efficiency and life span compared to brushed motors. Furthermore, brushless motor designs can have higher torque-to-weight ratios and can be controlled more accurately than brushed motors. Brushless DC motors are more expensive compared to brushed DC motors; nonetheless, they are a common choice for surgical robotic systems.

5.2 Piezoelectric Actuators

Although DC electromagnetic motors are the most common actuators utilized in surgical robotic platforms, **piezoelectric motors** are an alternative type of actuator which have seen deployment in neurosurgical robotic systems. Unlike electromagnetic motors, piezoelectric motors do not generate large magnetic fields which means they can be used in close proximity to magnetic resonance imaging (MRI) scanners. The NeuroArm robot utilizes piezoelectric motors so that interventional MRI procedures can be performed without causing any safety issues or image artifacts [2]. The piezoelectric effect is a phenomenon in which electric charge accumulates in some crystals and ceramic materials when subjected to a mechanical strain. The converse relationship is also true, when subjected to an electric field, these same materials will generate a mechanical strain. This latter physical principle can thus be exploited to create an actuator whereby an electric input can induce a controlled motion. In one of the simplest implementations, beam structures fabricated from materials that exhibit the piezoelectric effect will undergo a deflection in proportion to the driving voltage. In one such design implementation known as a bimorph; two piezoelectric layers can be independently strained to allow for bidirectional bending of a beam structure. These actuators can achieve high accelerations as they can be made very compact and lightweight. Piezoelectric bimorphs are an ideal actuator choice for handheld tremor suppression devices in which the actuation system needs to move as fast as the tremulous motion it is intended to compensate [11].

Since piezoelectric materials can only produce finite, and often very small strains when subject to an electric field; they must be integrated into a mechanism to form motors that can generate the larger-scale continuous motions required in many surgical robotic systems. Piezoelectric motors typically work by mechanically oscillating a piezoelectric material and transforming these small oscillatory motions into a continuous unidirectional motion. Continuous motion can be generated using a stepping approach analogous to how an inchworm locomotes. To understand this process, we must consider oscillation of the piezoelectric material in four phases. Initially, the piezoelectric material is in contact with a moveable component, such as a disk or translation stage, before deforming in such a way as to mechanically translate the moveable component a in a single, discrete step. Secondly, the piezoelectric material deforms in order to disengage from the moveable component. Thirdly, the material must then recoil in the reverse direction while remaining disengaged from the moveable component. Finally, the piezoelectric material can return to its initial stage and reengage the moveable component before repeating the translation phase. As this process repeats, this stepping sequence creates a continuous output motion of the moveable component. A common method for achieving this type of oscillatory motion is to generate standing waves in a single beam structure constructed from piezoelectric material. It is possible to excite a beam structure longitudinally in order to provide the repeated engagement and disengagement with the moveable component. At the same time, the beam structure can be excited in a bending mode to generate the traversing motions. When superimposed together, a wave-like excitation will generate the stepping actuation required to translate the moving component, this is illustrated in Fig. 7. The movable component may be a rotating disk or a translating component which can be used to drive a robotic joint.

5.3 Mechanisms

A mechanism is a device that transforms a force or motion input into a desired force or motion output. Mechanisms such as gears, belts, leadscrews and linkages are used in robotic systems to convert the motion generated by the robot actuators into motion at the robot joints. Here we focus on some of the basic mechanisms adopted in commercial neurosurgical robots (see Fig. 8).

Gears are ubiquitously adopted in a wide range of robotic manipulators. Gears are rotating disks with intermeshing teeth that enable power transmission: they can be configured to convert the speed, torque, or direction from a driving gear to a driven gear.

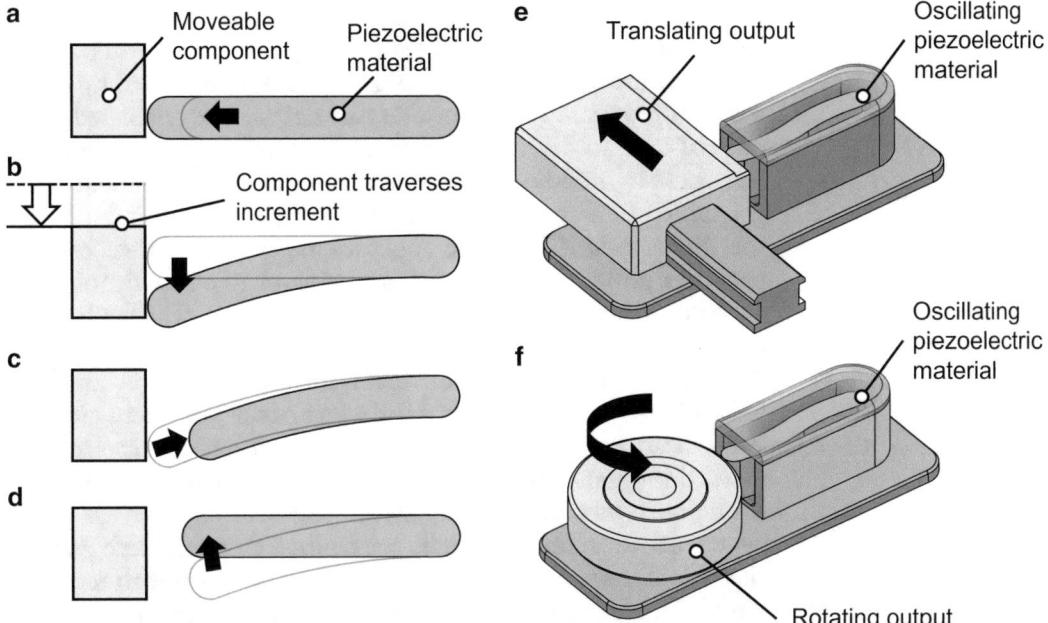

Fig. 7 The actuation mechanism for a piezoelectric actuator. (**a**) Elongation of the piezo structure that engages a moveable component, (**b**) bending of the piezoelectric structure to traverse the moveable component, (**c**) disengagement from the moveable structure, and (**d**) recoil to the initial state. This principle can be applied to a translating moveable component (**e**) or a rotating moveable component (**f**)

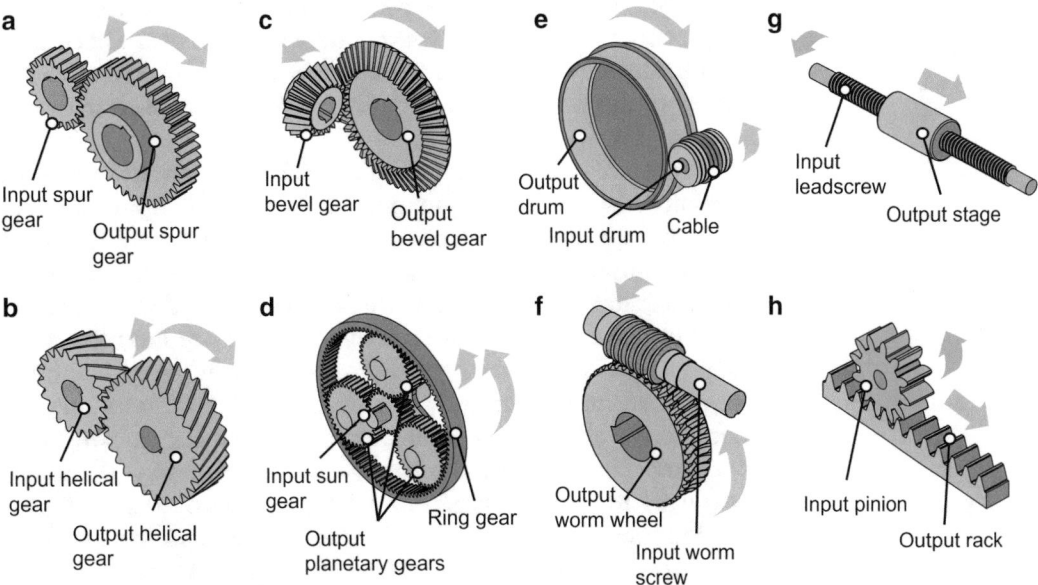

Fig. 8 Examples of basic mechanisms: (**a**) spur gear train, (**b**) helical gear train, (**c**) bevel gears, (**d**) planetary (or epicyclic) gear stage, (**e**) pretensioned capstan–cable assembly, (**f**) worm drive, (**g**) leadscrew assembly, and (**h**) rack -and- pinion. All mechanisms take a rotary input; those illustrated in green provide a rotary output and those illustrated in blue generate a translating output

Gears are commonly configured to create a mechanical advantage, increasing output torque and reducing the output speed. The speed change that occurs between two gears can be determined by selecting an appropriate gear ratio. The gear ratio can be calculated as the ratio of teeth in the driven gear relative to that of the driving gear. For example, a 20-tooth gear that drives a 30-tooth gear will have a gear ratio of 1.5 so that the driving gear will rotate $\times 1.5$ faster than the driven gear. In robotic applications, gears are often used to provide a power train from a rotary actuator to a robotic joint.

Spur gears are the simplest form of gear in which the teeth radially protrude from an inner hub and are oriented to be parallel with the axis of rotation. Gear teeth can also be manufactured in a helical form in such a way that the edges of the gear teeth are at an angle relative to the axis of rotation. Such helical gears can be meshed in parallel like spur gears or crossed so that the gears have orthogonal axes of rotation. Bevel gears are another means by which rotational motion can be transmitted through some nonparallel axis. These gears have a truncated conical form with teeth cut into the conical face which can mesh with another bevel gear. Gears are typically used to convert one form of rotary motion to another, but if a linear output motion is required then a rack-and-pinion assembly can be used. A pinion (gear) can be used to translate a toothed bar known as a rack which will translate as the pinion rotates. Worm gears are another type of gear arrangement that can be used to create large mechanical advantages within a single gear stage. Worm drives are comprised of a worm screw (a gear with a single tooth formed into a helical profile) and a worm wheel, with radial-protruding teeth. The worm screw drive axis is oriented at a tangent to the worm wheel so that the screw mechanism transmits power to the worm wheel.

Spur gears can be arranged as a serial drive train of individual gears mounted on bearing-supported shafts within a housing. An alternative approach often adopted in robotic systems is to use a planetary gearbox. In a planetary gearbox, there is a central "sun" gear which meshes with at least three equally spaced gears referred to as "planetary" gears. These planetary gears are mounted on a component called a carrier which is concentrically aligned with the sun gear. When the sun gear rotates, the planetary gears simultaneous rotate about their individual axes while also moving in a concentric path around the sun gear. The planetary gears also mesh with an outer ring gear which is also concentrically aligned with the sun gear. Planetary gear stages can be stacked to create a high-power density gearbox in which high gear ratios can be achieved within a relatively compact space.

A challenge inherent in many power transmission mechanisms is that of **backlash**. Mechanisms that have independent moving components such as gear trains must incorporate at least a small gap between each component. In engineering, this gap is referred to as a clearance. In gears, backlash is the lost motion that occurs as a result of the clearance between the meshing gear teeth. In many applications backlash is not a significant problem, but in robotic applications backlash can reduce the overall precision of the robotic manipulator. Gears can be machined to a high precision which can help minimize backlash, albeit at increased manufacturing costs. Another parameter of interest in mechanism design for medical robots is **backdrivability**. This is the ability of a mechanical system to be operated in the reverse direction to that which is primarily intended. In some surgical robotic applications, backdrivability is a desirable design requirement. For example, if power is suddenly lost during a surgery, the backdrivability of the robotic system means that it can be safely removed from the patient. Gear trains can be made backdrivable if the gear ratio remains relatively low, but higher gear ratio trains are nonbackdrivable due to the friction and inertia of the gearbox assembly.

Given the challenges of backlash and backdrivability with gear-based mechanisms, some medical robotic systems adopt cable drive assemblies to transmit power. Steel cables can be wrapped around **pulleys** or cylindrical drums and tensioned through a **capstan** assembly. Rotation of the capstan or pulley generates tension in the cable so that loads can be transmitted through the cable. Like gear trains, pulley systems can be configured to provide a mechanical advantage. Such mechanisms can be utilized in robot user interfaces or in surgical end-effectors. For example, the da Vinci® Endowrist® devices (Intuitive Surgical Inc., Sunnyvale, USA) are miniaturized robotic end effectors that mimic the human wrist. These instruments use a system of miniature pulleys and steel cables to transmit forces from the instrument shaft to the wristed instrument tip. In addition to cable drives, mechanical belts are another means by which power can be transmitted between pulley assemblies in an efficient and cost-effective manner.

Leadscrews are another type of mechanism commonly used in robotic applications. Leadscrews convert rotary motion into linear motion. Leadscrews are often used in robots that feature prismatic joints and adopt rotary DC electromagnetic motors for actuation. A leadscrew mechanism is based on the same principle as a nut and bolt wherein a helical thread converts a rotary input into a linear translational motion. A leadscrew mechanism is comprised of a threaded shaft that is coupled to a rotational actuator and supported in bearings so that the shaft can rotate. A female-threaded translation stage is mounted on to this threaded shaft and constrained so that rotation of the shaft causes the translation stage to traverse. A variant of the leadscrew is a ball screw which incorporates ball bearings within the transmission assembly to minimize friction. This type of mechanism is used to actuate the prismatic joints of the SpineAssist® system (Medtronic plc, Minneapolis, USA).

Linkages are another means by which different types of input motion can be converted to another form of output motion. Although linkages can be designed in three dimensions, planar two-dimensional linkages are more commonly adopted in practical applications. The simplest planar linkage is known as a four-bar linkage, wherein four bars are pinned together in a loop. Further bars can be added to the linkage chain if greater complexity output motion is required from the linkage. If the desired output motion profile is known, a linkage design can be synthesized to generate this desired motion profile for a given input motion. The input motion to a linkage will typically be a rotation or linear translation that can be generated by an actuator. An example of a linkage specifically adopted in some surgical robot designs is a remote center-of-motion mechanism. In keyhole interventions, it is desirable to be able to pivot a surgical tool about a virtual incision or access point. A linkage with bars organized in a parallelogram structure can achieve this type of motion and was adopted on the first generation of da Vinci® surgical robots (Intuitive Surgical Inc., Sunnyvale, USA). Should readers wish to garner a more comprehensive understanding of actuators and mechanisms, we recommend the following textbook chapter:

- Scheinman V, McCarthy J (2016) Mechanisms and actuation. In: Siciliano B, Khatib O (eds) Springer handbook of robotics. Springer, New York, pp 67–86.

6 Sensors

Like humans, robotic systems must exhibit the ability to sense both themselves (proprioception) and the external world (exteroception) in order to move in a controlled manner. While actuation systems fundamentally facilitate robot motion, sensing systems are critical to ensuring that these motions are controlled safely and precisely. Sensors are devices that can measure physical properties such as position, orientation, force, or temperature. Electronic-based sensors can be integrated into medical robotic systems in several ways. A key example is the use of position sensors in the joints of a robotic manipulator: these proprioceptive sensors facilitate feedback control so that the robot can precisely follow a desired motion trajectory. Other medical robotic systems incorporate exteroceptive sensing systems, for example force sensors to sense the delicate surgical manipulation forces experienced at the instrument tip. This type of sensor information can be used directly to control actuators within the robot or it might instead be relayed back to the surgeon, for example through a graphic or haptic-based user interface. Sensor information can also be used to inform high-level decision-making within the robotic control system, for example as an input trigger to execute a pre-programmed motion command. Sensors can also be used within robot safety systems, for example temperature sensors that detect motor overheating.

Although there are many sensing technologies available for integration into robots, in this chapter we will discuss two key exemplars which are ubiquitous in commercial neurosurgical robotic systems. The first sensor example we introduce is the rotary encoder. Rotary encoders are proprioceptive sensing devices designed to measure rotation and as such are commonly used to sense the angular position of motor-actuated robotic joints. The second example we will consider is the strain gauge, a device capable of measuring small deformations of mechanical structures which can be used to infer forces. Through these two examples, we will introduce some fundamental sensor concepts and terminology.

6.1   Rotary Encoders

Rotary encoders are electromechanical sensing subsystems capable of determining the angular motion of a shaft, typically the shaft of an electromagnetic motor. These devices are ubiquitous in robotic platforms across a broad spectrum of applications, as they provide proprioceptive positional sensing for robot manipulators. Rotary encoders can be broadly divided into two categories. Firstly, absolute encoders that can encode the absolute position of the motor shaft with respect to a fixed reference position within the encoder. The second category, incremental encoders can measure angular position relative to an arbitary reference point which can be defined using an initial homing procedure. Both types of rotary encoder typically incorporate a disk that is coupled to the driving actuator shaft and commonly adopt either optical or magnetic sensing technologies to detect angular motion in the disk. In an incremental optical-based encoder, the rotating disk is fabricated from an opaque material and incorporates equally spaced slot features that are formed in a concentric pattern about the disk axis (*see* Fig. 9a). A light source is positioned on one side of the disk and a light detector positioned on the opposing side at the radial location of the slot features. As the disk rotates, the slot features sequentially occlude and permit the passage of light to the detectors, creating a series of light pulses. Since the angular spacing of the slot features is known, a microprocessor can count the number of pulses to determine how far the shaft has rotated. Furthermore, the rate at which the light pulses are received by the detector determines the angular speed of the shaft. However, to interpret the direction that the encoder has rotated, a second light detector is required to count pulses at an angular phase shift relative to the first light detector. Comparison of these separate pulsed waveforms can then allow for the rotational direction to be determined (*see* Fig. 9b).

As with any sensor or sensing system, there are limits to the smallest change that can be detected. This limit is known as the sensor **resolution**. In the case of an optical encoder, the resolution

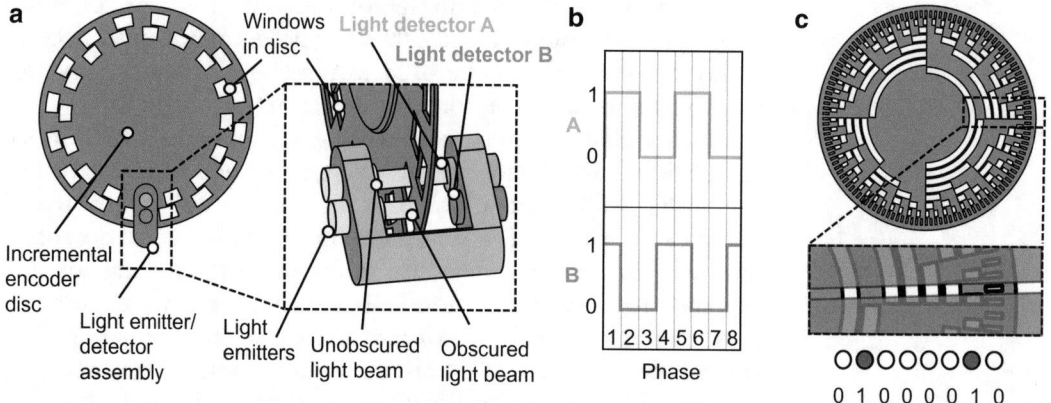

Fig. 9 (**a**) Example of a rotary incremental encoder assembly based on optical sensing. (**b**) The output generated by the two light detector systems generates a quadrature sequence which can be decoded to calculate rotation angle, speed and direction. (**c**) Illustrates an absolute encoder design in which each angle increment can be uniquely encoded

is limited by the angular slot spacing which determines the number of pulses per revolution (PPR). Typical encoders can range in resolution from 32-1024PPR. When considering the positional resolution requirements of the robotic system, it is also key to consider additional gearing in the actuation system. For example, a robotic joint actuated by a motor with a 64:1 gearbox and 1024PPR encoder can achieve a resolution of <0.006° (i.e., 360° ÷ (1024 × 64)) at the joint position. It is thus appreciable how electromechanical robotic systems can achieve such high precision motion using encoder-based technologies. An absolute optical encoder can also be fabricated by incorporating multiple concentric slot patterns about the disk axis in conjunction with a designated light detector for each slot pattern. For each angular position of the disk there is a unique optical pattern that can be decoded by a microprocessor and converted into the angle value (*see* Fig. 9c).

6.2 Strain Gauges

In surgical robotic systems, it is sometimes useful to sense the forces that the robot exerts on the patient during a procedure. Sensors that can measure the forces and torques during surgery commonly use devices known as **strain gauges** that can sense physical deformations of mechanical structures. All structures deform to a greater or lesser extent when subject to a mechanical force, and this deformation can be measured to infer both the magnitude and direction of force being applied to the structure (*see* Fig. 10). The term *strain* refers to a measure of deformation that a body undergoes under loading and is typically thought of as a ratio of the deformed state relative to the undeformed state. Strain gauges are electrical devices which use a resistance-based transduction principle to infer mechanical strain. These devices are typically constructed by

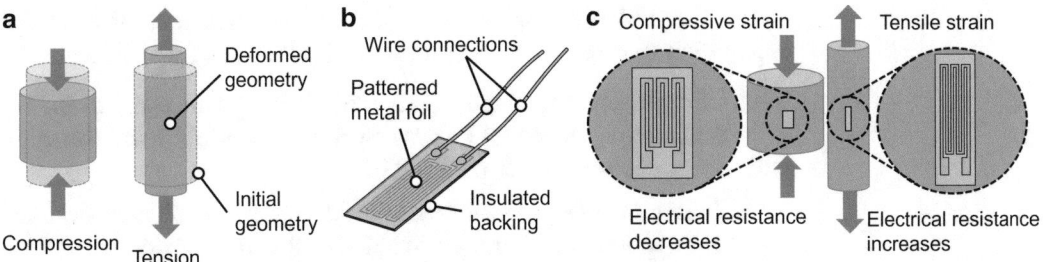

Fig. 10 (**a**) Illustration of how mechanical structures undergo geometrical changes when subjected to an external force, (**b**) features of a strain gauge, and (**c**) principle by which strain gauges operate

fabricating a thin metal foil into a zigzag pattern to create a structure that will undergo deformation. This electrically-conductive structure is adhered to an electrically-insulated backing and bonded to the mechanical structure of interest. When the geometry of the strain gauge foil structure is altered as a result of an external mechanical strain, a measurable change in electrical resistance can be observed. If the strain gauge undergoes a tensile strain, the foil structure will elongate which increases the path of the conductor and also reduces the cross-sectional area. The converse behavior occurs if the strain gauge undergoes a compressive strain. The resistance of the strain gauge is dependent on both the path length and cross-sectional area of the foil structure. Thus, a change in the strain gauge electrical resistance can be correlated with mechanical strain. Since this mechanical strain can be measured, a force can be inferred by analyzing the mechanical structure that has undergone deformation. Like many sensing elements, strain gauges must usually be combined with an electrical amplifier which converts a small change in electrical resistance to an electrical signal that can be accurately interpreted by a robot control system. If the strain gauge is integrated into a force sensor it can be calibrated against a known load in order to determine the sensor **sensitivity**. The sensitivity of a sensor (such as a strain gauge-based force sensor) is the ratio between the sensor output signal and the quantity being measured (i.e., with units Volts/Newton).

This section has reviewed two key sensor technologies commonly utilized in robotics. Many other additional sensing technologies can be integrated in to robotic systems. For example optical imaging devices, such as cameras, can be used to provide information about the environment in the robot coordinate system. Similarly, optical tracking systems can provide an accurate pose of the robotic system and how it relates to other tracked markers that might be placed on other tools or the patient. Furthermore, medical imaging modalities such as ultrasound, fluoroscopy and MRI, can be utilized as input sensors to actively control a medical robotic system.

Should readers wish to garner a more comprehensive under-standing of sensors, we recommend the following text books:

- Christensen HI, Hager G (2016) Sensing and estimation. In: Siciliano B, Khatib O (eds) Springer handbook of robotics. Springer, New York, pp 87–107
- Fraden J (2004) Handbook of modern sensors: physics, designs, and applications. Springer Science & Business Media, New York

7   Control

We have reviewed the actuators and mechanisms that create motion in robots and the sensing systems that enable exteroception and proprioception. Control systems are the integration of these various components that enable a robot to execute precisely controlled motion. In engineering, the term *control* is used to describe the technology by which a process or procedure is performed, or in other words the way in which a piece of equipment is manipulated. All types of machinery, from boilers and ovens to heavy industrial robots in factories, use some sort of control in order to be functional. Actuators, mechanisms, and sensors are all components that transform some form of input to an output. For example, a DC motor requires an electrical current input and produces an output torque at the rotor shaft (Fig. 11 provides some additional examples of such device processes). A wide range of engineering components can be theoretically modeled so that they can be analyzed in a control system. The theoretical model of a component, specifically its output behavior in relation to any given input, is referred to as the **transfer function**. These individual components can be configured together into more complex systems that can be controlled. The goal of the control system is to select an appropriate system input in order to generate a desired future system output.

There are two main categories of control: open-loop systems and closed-loop systems. In an open-loop control system, the only variable that is being considered when deciding what action is going to take place is the input. Figure 12 illustrates an open-loop control system for actuating a robotic joint. In control theory, the term **plant** refers to the system which is to be controlled. In this case, the plant is a chain of components: a DC motor, a gearbox, and a revolute joint. In this system, the input is an electrical current which is supplied to the DC motor, this motor then generates a torque at the output motor shaft. The motor shaft is coupled to a gearbox which amplifies the torque, according to the gear ratio. Finally, the torque applied at the robot joint causes rotation of the robot arm; the angle through which the joint rotates is governed by the laws of motion. Each of these processes (by which an input is

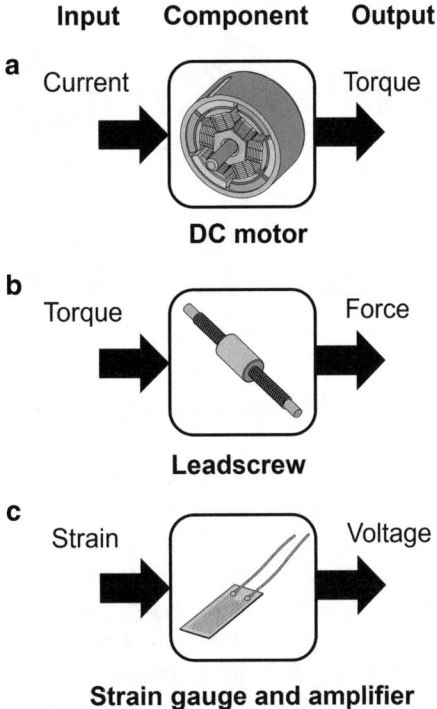

Fig. 11 Physical processes from different devices transform an input into an output. (**a**) A DC motor is supplied with a current and provides an output torque, (**b**) a leadscrew is a mechanism that converts a torque load into a linear force, and (**c**) a strain gauge (with amplifier) is an example of a sensor which provides a voltage output in response to a strain input

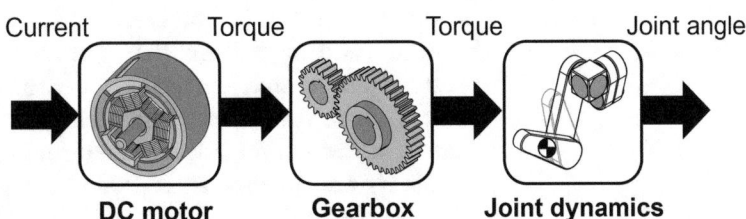

Fig. 12 A chain of component processes configured into an open-loop system for actuating a robotic joint

transformed to an output) can be modeled using physical principles and engineering methods. It is thus possible to predict the overall system output (movement of the robot joint) based on the system input (current supplied to the DC motor). However, one problem with this approach is that it assumes perfect prediction of each individual process. Furthermore, this control methodology cannot account for any external factors that might alter the overall system output.

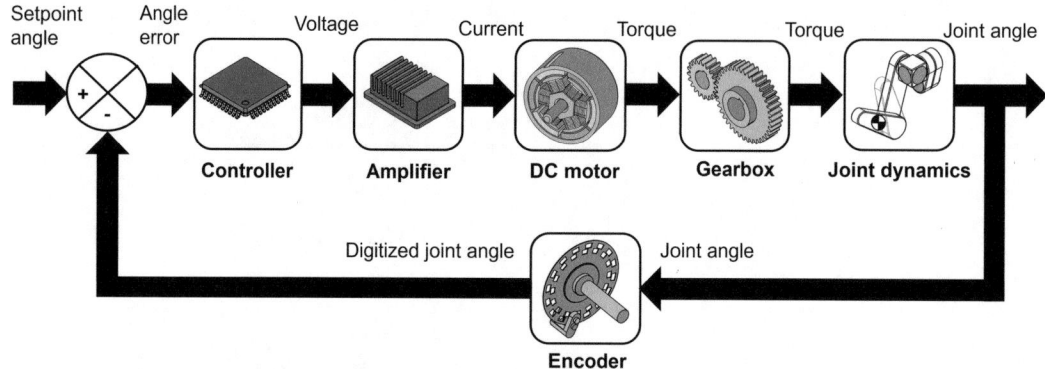

Fig. 13 A closed-loop control system for actuating a robotic joint

In a closed-loop control system, the system output influences the control action on the plant. For example, in controlling a robotic joint angle, the current state of the joint angle (the output) will influence how much current is supplied to the DC motor. Such a methodology requires the incorporation of a sensing system that can monitor the system output, such as an encoder to measure the robot joint position. In a closed-loop control system, another necessary system input is the **setpoint** (also known as the *reference*). The system setpoint is the desired state of the system, such as the desired angle of the robot joint. This setpoint is compared to the system output, as measured by the sensing system, in a continuous loop. The discrepancy between the setpoint input and the system output at any given instance in time is referred to as the **error**. It is this error value that forms the input to a control process which generates an appropriate output to control the plant. The control process is executed by the **controller**, which is the computational hardware and algorithms that generate an output signal to the plant. If configured correctly, the controller will command the plant to drive the system error to zero over time. Figure 13 illustrates a closed-loop control system for controlling a robotic joint. In this system, an encoder is used to measure the robot joint angle which is then compared to a reference angle (the setpoint). The controller reads in this error signal and generates an appropriate output voltage which is sent to an electrical amplifier. The amplifier generates the current required to actuate the plant (the DC motor, gearbox, and robot joint), which causes rotation of the joint. This process continually repeats in a loop so as to regulate the position of the robot joint according to the reference angle.

7.1 Proportional--Integral–Derivative Control

A very common type of closed-loop control algorithm is the **proportional–integral–derivative (PID)** controller. It is one of the most reliable and commonly adopted control methodologies used in robotic applications. As we have seen, the input to a controller is the error between the setpoint and the measured state of the system. The terms proportional, integral, and derivative refer to three parallel pathways in which this error value is processed by the controller. The summation of these processed error terms forms the output generated by the controller.

Proportional control refers to a methodology for varying the controller output in proportion to the controller input. The error term is simply multiplied by a factor known as the proportional gain. Thus, if the control system detects a large error, the output to correct this error will be proportionally large so as to quickly reduce the error. Once the system has reached the setpoint and there is no error, the controller output will also be zero. The use of proportional control alone has limitations. Firstly, control systems can **overshoot** in trying to reach the desired setpoint. For example, a robotic arm may be supplied with a surge of current in response to a new setpoint angle. The current supplied to the motor will diminish as the arm accelerates toward the setpoint angle, but the mechanical inertia of the robot joint will cause it to overshoot the setpoint angle. The controller will then detect a negative error and reverse the current supplied to the motor to rotate the joint back toward the desired position. This process will then repeat, and the robot arm will oscillate back and forth about the desired setpoint angle. The robot arm may then eventually settle after a period of time (the **settling time**), or it may oscillate indefinitely. One solution to this problem is to reduce the proportional gain so that the system does not overshoot so profoundly and settles relatively quickly. However, this will increase the **rise time** of the control system; it will take longer for the robotic joint to approach the setpoint angle since the controller output is reduced and less current is supplied to the motor. The step response of a robot arm using closed-loop PID control is graphically represented in Figs. 14 and 15.

To combat the overshooting problem, the proportional control path can be combined with a derivative control path to create a proportional–derivative (PD) controller. The purpose of this control scheme is to better compensate for future events. Rather than just interpreting the error term alone, the derivative pathway also considers the rate at which the error is changing over time and adjusts the controller output accordingly. The rate-of-error change is multiplied by another factor, the derivative gain, and the proportional and derivative terms are summed together to create the controller output. The derivative term attenuates the controller output if the error is reducing rapidly, effectively acting like a preemptive brake to minimize or prevent the system overshooting the setpoint. Figure 15b illustrates how a PD controller can attenuate overshooting.

A PD controller will effectively command a robot arm to the desired position; however, a small **steady-state error** will still remain. Once the system has settled so that there is no change in error rate, the derivative term will contribute nothing to the

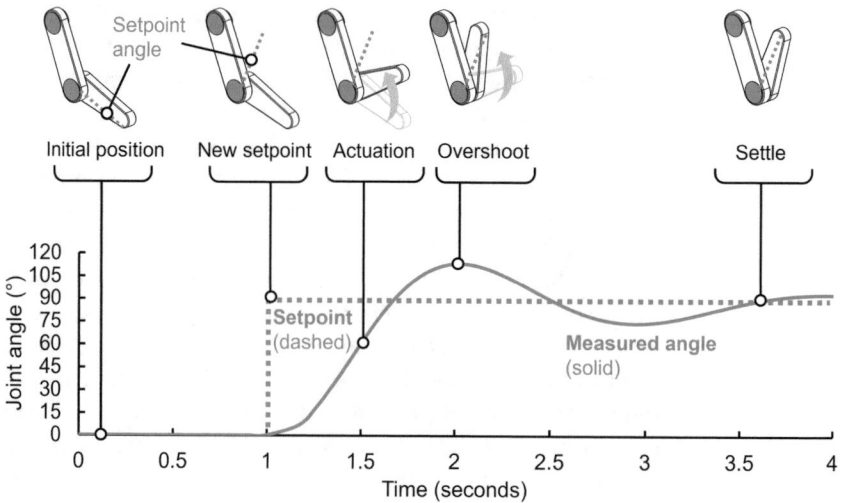

Fig. 14 Graphical illustration of how a PID controller can manipulate a robotic joint angle when subject to a step change in the setpoint

controller output. The proportional path of the controller will still generate an output in proportion to the residual error, but since this error is very low, the controller output will be correspondingly low. This low controller output (the input to the plant) will often be insufficient to overcome the final residual error. For example, in a robot joint, a low controller output corresponds to a low current that is supplied to the motor. The low torque generated by the motor will be insufficient to overcome any residual friction or inertia. Increasing the proportional gain will help reduce the steady-state error, but as we have discussed, this solution creates other problems such as overshooting. An alternative solution is to incorporate the third control term: integral control. Whereas derivative control is concerned with compensating future events, integral control helps the controller compensate events of the past. The error term can go through an integration process in which all of the past errors are summed together with every iteration of the control loop. This accumulation of error is then multiplied by a third gain, the integral gain, and added to the proportional and derivative terms, so as to create the complete PID control output. If a steady-state error exists in the system, while initially the accumulated error will be very small, over a period time the accumulated error will increase. As the accumulated error increases, the controller output will increase in order to drive the error to zero. Figure 15c illustrates how the introduction of integral control can eliminate steady-state errors.

PID control is a very effective means of controlling dynamic systems such as robots. Enabling robotic joints to execute tasks repeatably, while rejecting external disturbances that would otherwise perturb the system in performing its task. PID controllers are,

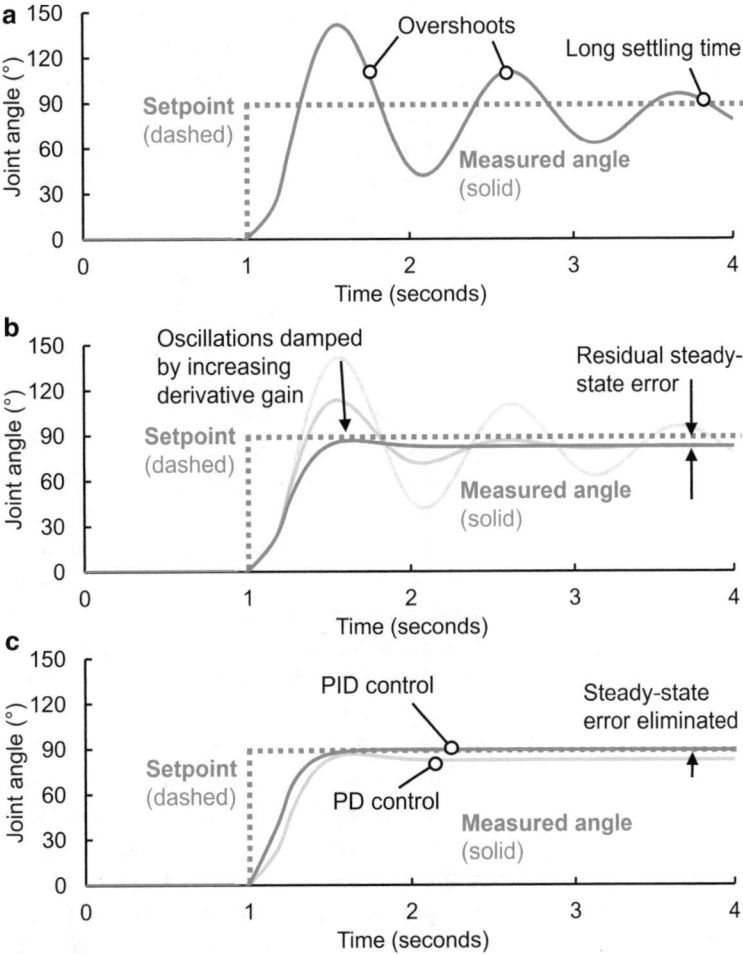

Fig. 15 Illustrative closed-loop position control responses using PID controllers. (**a**) Shows how simple proportional control can generate overshoots and oscillatory behavior within the control system. (**b**) Shows a step response using a PD controller, increasing the derivative gain reduces and eliminates overshoots. (**c**) Shows how the introduction of integral control can eliminate steady-state errors to maximize the accuracy of the control system

nonetheless, only effective if the correct proportional, integral and derivative gains have been selected. In control engineering, there is a trade-off between system performance and system **stability**. If the controller gains are set too high, the system can become unstable and the robot will behave erratically. Control engineers will typically tune a robot controller so as to exhibit acceptable system performance without any risk of becoming unstable. It is possible to tune the controller gains to meet different performance requirements. For example, one set of controller gain values might enable a fast rise time but result in some overshooting. Another set of controller gain values might ensure no overshooting but at the cost of an increased system rise time.

The two main methods used to control a robot are position and force control. In position control, the main task that needs solving is the execution of desired motions in free space, such as the example of the robot joint illustrated in Fig. 13. In this example, we studied the control of a single DoF robotic joint angle using a rotary encoder as the feedback sensor. For a robotic system with multiple joints, multiple independent control loops must be used to manipulate each joint of the robot. While these low-level controllers manage the motion of each joint, a higher-level controller must compute the coordinated motion of the entire robotic system. Calculation of the inverse kinematics will yield the required joint positions which can be relayed to each individual joint-level control loop. Many neurosurgical robotic systems use this type of position control system, such as the SpineAssist® device (Medtronic plc, Minneapolis, USA) which is manipulated in real-time based on predefined computed tomography images.

In other circumstances, it is desirable to control the force that a robot end-effector is exerting on the environment (such a surgical tool applying forces to the brain). In this case we must use a force control scheme. For a robot manipulation task to be successful, it is essential that the interaction between the manipulator and the environment is smooth, especially in surgical applications such as neurosurgery, where even relatively low forces can cause iatrogenic injury to the patient. In contrast to position control, where each joint reaches a target angle or position, in force control the output torque is controlled to match the desired force applied by the interaction of the end-effector with its environment. Force or torque sensors are typically used to measure the external force that is being applied to the end-effector and this value is used to close the loop and produce the corresponding joint torques to match the external force. Force control schemes can be used to enable haptic feedback which is the simulated touch between a robot and a human user. Some surgical robotic systems, such as the NeuroArm [2], are capable of relaying the surgical forces exerted on the surgical robot back to the operating surgeon via teleoperation. An alternative approach has been to develop hands-on instruments with enhanced force-feedback functionalities [12, 13]. This chapter gives an overview of basic control systems, should readers wish to garner a more comprehensive understanding of control systems, we recommend the following textbook chapters:

- Chung W, Fu L, Hsu S (2016) Motion control. In: Siciliano B, Khatib O (eds) Springer handbook of robotics. Springer, New York, pp 133–159

- Villani L, De Schutter J (2016) Force control. In: Siciliano B, Khatib O (eds) Springer handbook of robotics. Springer, New York, pp 161–185

8   User-Interface Paradigms

This chapter has reviewed some of the technical fundamentals required to construct a robotic system. In this final section, we introduce the user-interface paradigms by which a neurosurgeon may operate a robotic system to perform a surgical task. Medical robotic systems can be broadly stratified into three categories: supervisory-control robots, teleoperated systems, and shared-control robots.

8.1   Supervisory-Control Robots

Supervisory-control robotic systems are largely automated systems that bear significant resemblance to conventional industrial robots. A supervisory-control robot is preprogrammed to execute a specific set of tasks in order to perform a surgical intervention. Such systems require considerable preoperative planning for the robot to execute its tasks effectively since supervisory-control robots cannot perform readjustment themselves during an intervention. This robot paradigm is often used in conjunction with preoperative medical imaging and image guidance systems. **Coregistration** of 3D medical imaging data with the patient allows a robotic manipulator to plan how an intervention is performed. Needle placement procedures and radiotherapy are typical applications for supervisory-control systems, since positional precision can be greatly enhanced with a robotic manipulator. This type of robot has seen relatively significant uptake in brain and spine interventions because coregistration of the cranial and spinal anatomy can be performed accurately since these structures are relatively rigid.

8.2   Teleoperated Robots

Teleoperated robotic systems allow a surgeon to remotely control an interventional robotic device in real-time. The surgeon typically sits at a console in which they are provided with a view of the surgical scene. At the console, the surgeon can manipulate linkage mechanisms which track the surgeon's input motion. These motions are then processed by a computer control system and are replicated on a robot that performs the intervention on the patient. This type of robotic system is refered to as being "teleoperated" because the surgeon does not directly operate on the patient, but instead through a "fly-by-wire" system. This user-interface approach can be advantageous since the robot system can operate with high precision, but executive control at any given instance remains with the surgeon. The teleoperated robotic system paradigm has seen significant clinical uptake in the wider field of medical robotics; notably the da Vinci® robot (Intuitive Surgical, Sunnyvale, CA, USA) has been heavily adopted in several clinical applications. Teleoperated systems can also allow for surgeries to be performed in physically challenging environments. For example, a significant advantage of the NeuroArm system is that it can perform surgery in very close proximity to an MRI scanner, allowing a surgeon to sequentially intervene and rescan a patient to ensure a full tumor resection. Teleoperated robots can also be advantageous in allowing surgeons to operate at smaller physical scales since the interventional robotic tools can be miniaturized and their motion scaled down. Teleoperated systems are thus potentially advantageous in performing keyhole neurosurgical interventions that are otherwise challenging to carry out manually.

8.3 Shared-Control Robots

Shared-control robots are designed to allow a surgeon to directly control the robotic platform in real time and retain executive control. Unlike teleoperated systems, a shared-control robot can be directly manipulated by the operating surgeon rather than through a remote console. This greatly simplifies the complexity of the operating robot by removing the necessity for a remote console and the associated hardware. Like teleoperated systems, shared-control robots fuse together the respective advantages of humans and robots. Shared-control systems allow interventions to be carried out with superhuman precision or delicateness, but high-level judgment and decision-making remains with the surgeon. Shared-control robots can be anchored to the ground like a conventional robotic manipulator, or alternatively, robotic features can be integrated into ungrounded handheld devices [12]. This latter class of "smart," handheld tools can offer some of the advantages of larger robots but with lower costs, and the potential for seamless integration into the surgical workflow.

Glossary

Actuator – A device capable of converting an input energy source into an output motion.

Armature – In the context of an electromagnetic motor, the armature is a component that contains electrical windings which generate an oscillating electromagnetic field when supplied with an alternating current.

Base – The robot base is typically the point at which a robot is anchored to a reference frame, such as the ground. Kinematic calculations are usually performed with reference to the robot base.

Backdrivability – The ability of a mechanism, such as a gear train, to operate in reverse.

Backlash – (Also referred to as "play") is the small clearance between rigid body mechanisms, such as gears, which causes lost motion and positioning inaccuracies in robotic systems.

Brushed motor – Electromagnetic motor that uses brushes and a commutator to mechanically switch a direct electrical current to perpetuate rotary motion in response to a direct current input.

Brushless motor – Electromagnetic motor which is electrically commutated so that individual windings are selectively energized in a sequential manner to perpetuate unidirectional rotary motion.

Capstan – Rotating drum in which a cable is wrapped around for the purpose of power transmission.

Coregistration – The process by which a digital, 3D anatomical model based on preoperative imaging of a patient is aligned with the patient's actual geometry during an image-guided intervention.

Commutator – In the context of a direct current electrical motor, the commutator is a rotary mechanical switching mechanism that switches the polarity of current supplied to the armature so as to perpetuate rotary motion.

Continuum robot – A robotic system that achieves motion by physical deformation rather than through discrete motion of individual rigid bodies, continuum robots are commonly configured in tentacle-like configurations for minimally invasive surgical applications.

Controller – In robotics, a controller is the computational hardware and software algorithm that processes sensory input information and outputs actions to an actuation system.

Degree-of-freedom (DoF) – In robotics, a degree-of-freedom refers to an independent mode by which a robot can move.

Dexterous workspace – Dexterous workspace refers to a volume within the reachable workspace of a robotic system in which the robot end-effector can reach in any orientation.

Direct current (motor) – Direct current motors are actuators that are activated by a unidirectional electrical current, as opposed to an alternating current input.

Dynamics – In general terms, dynamics refers to a subfield of mechanics that relates how forces acting on a body cause motion. In robotics, dynamics specifically refers to the relationship between forces and motion in a robotic system.

Electromagnetic motor – An actuator that use the principles of electromagnetism to achieve motion, usually through a combination of permanent magnets and electromagnets.

End-effector – In a robot, the end-effector refers to the distal feature of a robot arm or system, in the context of a surgical robot this is typically a surgical tool or imaging device.

Error – In control engineering, error refers to the discrepancy between the system setpoint (the desired state of the control system) and the actual, measured state of the system.

Error (steady-state) – In control engineering, a steady-state error is a residual error that does not attenuate over time.

Forward dynamics – A process of computing a robot's motion given the known input forces and torques at the robot joints.

Forward kinematics – A process of computing a robot's motion given the known input motions of each joint in the robot.

Gear – A tooth-wheeled mechanism that can be meshed with at least one other gear, where one gear is driven and the other gear is driving, so as to facilitate the transmission of power.

Inverse dynamics – A process of computing the forces and torques required at a robot's joints to generate a predefined motion in a robot.

Inverse kinematics – A process of computing the motions required at a robot's joints to generate a predefined motion in a robot.

Joint – A robotic joint refers to the moveable components in a robot that are coupled together via links, joints can be revolute or prismatic.

Kinematics – In general, kinematics refers to a subfield of mechanics which studies the motion of bodies without concern for the forces that cause motion. In robotics, kinematics specifically refers to calculations that relate the motion of a robot to the motion of its individual joints.

Leadscrew – A leadscrew is a mechanism that converts rotary motion to linear motion using a helical thread.

Link/linkage – A robot link is the rigid body that connects robotic joints to form a multibody system. Similarly, linkage mechanisms allow for power transmission from an input link to an output link.

Overshoot – In control engineering, the term overshoot refers to the tendency of a control system to exceed its target setpoint.

Parallel robot – A robotic system in which multiple joint-linkage chains are coupled to the same end-effector.

Piezoelectric motors – Actuators that generate motion based on the piezoelectric principle.

Plant – In control engineering, the plant refers to the system being controlled.

Prismatic joint – A moveable component in a robot with a translational degree-of-freedom.

Proportional–integral–derivative (PID) – In control engineering, PID refers to a control loop algorithm designed to drive a control system error to zero.

Pulleys – A rotary, wheel-based mechanism used in conjunction with a cable or belt for the purpose of power transmission.

Reachable workspace – Reachable workspace refers to the volume in which a robot end-effector can physically reach within.

Redundancy – When the number of degrees-of-freedom of a robotic manipulator is more than necessary to execute a specific task.

Resolution – In the context of sensors, resolution refers to the smallest change that the sensor can detect in the quantity being measured.

Revolute joint – A moveable component in a robot with a rotational degree-of-freedom.

Rigid robotic system – A robotic system in which motion is created through rigid motions, i.e. rotations and translations, rather than deformations. In rigid body robotic systems, kinematic and dynamics calculations are performed on the assumption of perfectly rigid joints and linkages.

Rise time – In control engineering, the rise time refers to the time period required for a control system to initially reach a new target position, typically in reference to a step change input.

Rotary encoders – Electromechanical device capable of measuring the angular position of a shaft by encoding an angular position into an electrical signal.

Rotor – In the context of an electromagnetic motor, a rotor is the component within the motor that moves and is coupled to an output shaft.

Sensitivity – In sensors, the extent to which a sensor's output signal will vary with respect to changes in the quantity being measured.

Sensor – A device that can measure physical properties such as: position, orientation, force or temperature by generating an output signal in relation to the input stimulus.

Serial robot – A type of robotic manipulator in which a series of joints and links are connected together in a single chain, from the robot base to the end-effector.

Setpoint – In control engineering, the setpoint (also known as the reference) is the desired state of a system.

Settling time – In control engineering, the settling time is the time required for a control system to settle upon a new setpoint after a disturbance such as a step-change input.

Shared-control robotic system – Is a user-interface paradigm in surgical robotics by which a surgeon and robot share control of a surgical tool, so as to enhance surgical precision and allow the surgeon to retain executive control of the robot. Shared control robots can be grounded to a fixed reference frame or be handheld.

Singularity – A condition in which a robot cannot move in a desired direction regardless of how the joints move.

Stability – In control engineering, stability refers to a control system producing a bounded output for a given bounded input. In practical terms, a stable control system will behave in a predictable manner even when subject to disturbances.

Strain gauge – Sensing device that can measure mechanical strain using a resistive transduction principle.

Stator – In an electromagnetic motor, the stator is the sub-assembly of components that remain stationary.

Supervisory-control robotic system – A robotic system that executes tasks autonomously.

Teleoperated robotic system – A robotic system user-interface paradigm that allows a surgeon to operate on a patient remotely, via a console that can be used to control an interventional robot. Teleoperated robotic systems are also sometimes refered to as master-slave robotic systems.

Transfer function – The theoretical model of a component which captures a device's output for any given input.

References

1. Siciliano B, Khatib O (2016) Springer handbook of robotics. Springer, New York
2. Sutherland GR, McBeth PB, Louw DF (2003) NeuroArm: an MR compatible robot for microsurgery, International congress series. Elsevier, Amsterdam, pp 504–508
3. Burgner-Kahrs J, Rucker DC, Choset H (2015) Continuum robots for medical applications: a survey. IEEE Trans Robot 31:1261–1280
4. Camarillo DB, Milne CF, Carlson CR et al (2008) Mechanics modeling of tendon-driven continuum manipulators. IEEE Trans Robot 24:1262–1273
5. Simaan N (2005) Snake-like units using flexible backbones and actuation redundancy for enhanced miniaturization. In: Proceedings of the 2005 IEEE international conference on robotics and automation. IEEE, New York, pp 3012–3017
6. Webster RJ III, Romano JM, Cowan NJ (2008) Mechanics of precurved-tube continuum robots. IEEE Trans Robot 25:67–78
7. Dupont PE, Lock J, Itkowitz B et al (2009) Design and control of concentric-tube robots. IEEE Trans Robot 26:209–225
8. Webster RJ III, Kim JS, Cowan NJ et al (2006) Nonholonomic modeling of needle steering. Int J Robot Res 25:509–525
9. Minhas DS, Engh JA, Fenske MM et al (2007) Modeling of needle steering via duty-cycled spinning. In: 2007 29th Annual international conference of the IEEE Engineering in Medicine and Biology Society. IEEE, New York, pp 2756–2759
10. Ko SY, Frasson L, Baena FR (2011) Closed-loop planar motion control of a steerable probe with a "programmable bevel" inspired by nature. IEEE Trans Robot 27:970–983
11. MacLachlan RA, Becker BC, Tabarés JC et al (2011) Micron: an actively stabilized handheld tool for microsurgery. IEEE Trans Robot 28:195–212
12. Payne CJ, Yang GZ (2014) Hand-held medical robots. Ann Biomed Eng 42:1594–1605
13. Taylor R, Jensen P, Whitcomb L et al (1999) A steady-hand robotic system for microsurgical augmentation. Int J Robot Res 18:1201–1210

Chapter 2

Supervisory-Control Robots

Vani Virdyawan, Riccardo Secoli, Eloise Matheson, Marlene Pinzi, Thomas Watts, Stefano Galvan, and Ferdinando Rodriguez y Baena

Abstract

The supervisory-control method is used in the majority of neurosurgical robots to date where the surgeon makes the high-level decisions, which are then autonomously performed by the robot. In this chapter the differences in the roles of the robots during preoperative and intraoperative procedures are explained. During intraoperative procedures the robot can have either direct interaction or no direct interaction with the human tissues, called active and passive systems, respectively. The flow of information between the robots, the surgical environment, and the surgeons, to enable these forms of interaction, is also discussed. Examples of currently available robotic systems are provided.

Key words Neurosurgical robotics, Supervisory-control, Minimally invasive neurosurgery, Passive robotics system, Active robotic system, Preoperative planning

1 Introduction

Sheridan [1] defined supervisory-control as when "one or more human operators [set] initial conditions for intermittently adjusting and receiving information from a computer that itself closes an inner control loop through electromechanical sensors, effectors, and the task environment." In surgical robotics, this form of assistance has taken over a decade to develop from when the first robot was deployed in an operating theatre in the late 1980s [2] to the first use of robots to automatically remove human tissue in the late 1990s [3], and a further decade to become mainstream. In such systems, as in the original Robodoc platform for robotic-assisted hip and knee replacement surgery (now TSolution One®, THINK Surgical, Inc.), the surgeon makes high-level decisions, and the robot subsequently performs surgical tasks autonomously [4].

The information flow between the supervisor (human/surgeon) and the robot is shown in Fig. 1 [5]: the surgeon gives a task to the robot; the robot then performs the task by employing information from environmental sensors to close the control loop;

Hani J. Marcus and Christopher J. Payne (eds.), *Neurosurgical Robotics*, Neuromethods, vol. 162,
https://doi.org/10.1007/978-1-0716-0993-4_2, © Springer Science+Business Media, LLC, part of Springer Nature 2021

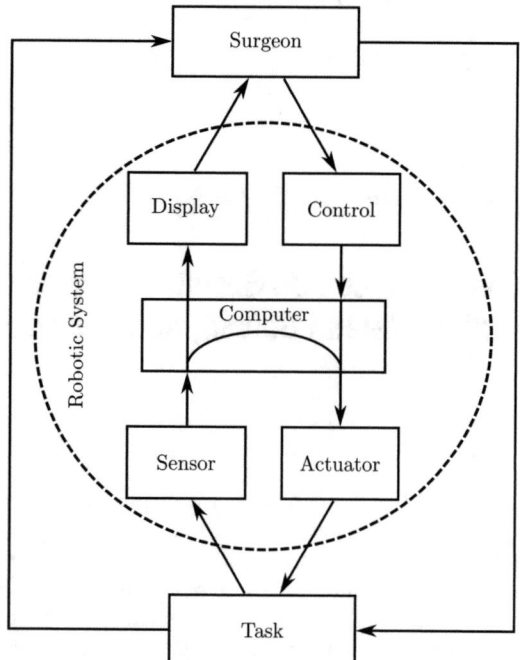

Fig. 1 The general model of supervisory-control system in surgical robotics. (Copyright ©2012 by John Wiley & Sons, Inc. Adapted with permission from [1])

the surgeon receives information of the process through the display of the robotic system or through the task directly; and the surgeon can update the task by updating their input to the robot or by manipulating the task space directly.

Using neurosurgical robotics as a useful case study, Joskowicz et al. [6] describes the surgical protocol for minimally invasive (MI) keyhole neurosurgery as consisting of preoperative and intraoperative procedures. The preoperative procedure consists of the following steps:

- Preimaging preparation: implant skull screws and/or attach skin markers.

- Image acquisition: acquire a CT/MRI image.

- Planning: generating preoperative planning based on the imaging results.

  The intraoperative procedure then consists of the following:

- Preparation: setup of the support system and entry-point incision.

- Localization: location of the MI instrument (catheter, needle, probe, biopsy tool, etc.) at the chosen entry point and orientation adjustment.

- Guidance: providing mechanical guidance for instrument insertion.

- Insertion: inserting the instrument to the planned depth at an appropriate speed and with appropriate force.

In this context, a surgical robot employing a supervisory-control modality can assist the surgeon in one or multiple steps, as described above.

Several examples of neurosurgical robotic systems exist in literature (*see* refs. [7, 8] for an exhaustive review), such as Neuromate [9], ROSA [10], Minerva [11, 12], MARS [6], Renaissance [13], MKM [14, 15] and our own EDEN2020 neurosurgical platform for MI interventions [16]. Virtually all of these systems are controlled using a supervisory-control method; hence, we will use this category to illustrate some of the variations available in today's state of the art. This chapter discusses neurosurgical tasks that can be performed based on a supervisory-control method using examples from a number of existing robotic technologies.

2 Preoperative Procedure

In the preoperative phase, the robotic system can assist the surgeon during the planning process by generating a suitable path to reach a target from computer tomography (CT) or magnetic resonance imaging (MRI) volume data. For instance, the NISS system [17] has a subsystem called NeuroPlan to carry out the path planning process. Using NeuroPlan, the surgeon first defines the areas that have to be avoided, such that the software can automatically generate the required path, which meets both anatomical and instrument-related constraints for a given procedure, such as size, depth, and geometry The surgeon then verifies the path before it is loaded in the NISS robot for intraoperative assistance.

The MARS robot system (originally by Joskowicz et al. 2006 [6] and now commercialized as the Renaissance System, by Medtronic Inc.), on the other hand, provides a module to automatically compute the optimal placement of the robot base, which is bone-mounted, and its range. The MARS system (Fig. 2 Left) includes a miniature robot that can be mounted either on the patient skull or on a head immobilization clamp. After choosing the target and entry points in the skull, the module then computes the optimum robot placement so that the motion of the robot tip guide from its home position is minimized. The module also provides alternative robot configurations in case the optimal placement is not clinically viable (Fig. 2 Right).

These two robotic system examples employ a rigid instrument to perform the procedure, limiting the planning choices available to the surgeon to a straight line. Steerable needle systems, which are

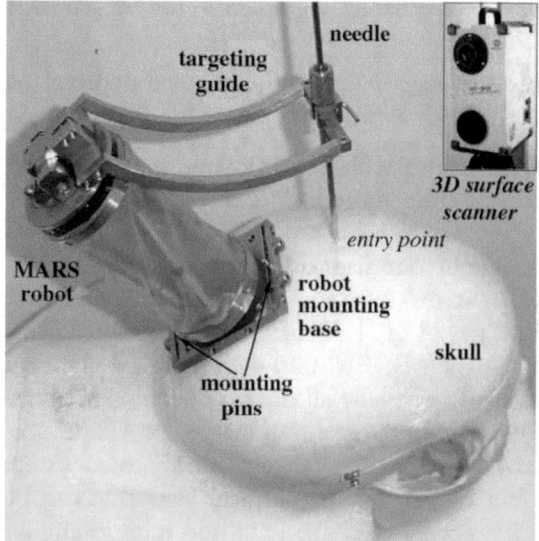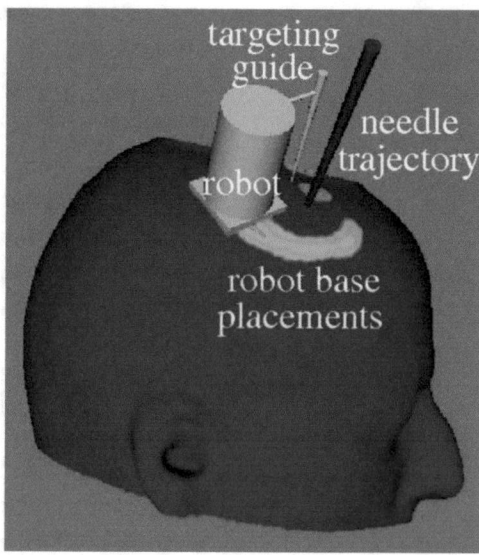

Fig. 2 Left: MARS robot mounted in the skull, Right: the optimum MARS placement in the skull generated by the planner module and its alternative placements. (Copyright ©2006 by Informa UK Ltd. Reprinted with permission from [6])

slowly emerging within the literature, can extend these to complex trajectories able to intersect deep lesions while avoiding obstacles. Indeed, steerable needles can follow a curvilinear path in the brain to circumvent critical structures and account for tissue deformation due to tool–tissue interactions, deliquoration, breathing, and pulsatile motion. Steering inside tissue is achieved by controlling the bending of the needle either as a direct result of passive needle–tissue interaction forces or via active modification of the needle shape [18].

One such steerable needle system has specifically been designed to access deep lesion inside the brain, and was engineered within our Mechatronics in Medicine Laboratory at Imperial College. The biologically inspired design, which takes inspiration from the egg-laying channel (ovipositor) or wood-boring wasps, takes the name of programmable bevel-tipped needle (PBN) [19]. Unlike conventional needles, PBNs possess a multisegment design (*see* Fig. 3), which enables the tip to steer in full three dimensional (3D) space thanks to asymmetric passive forces acting at the needle tip. More specifically, Watts et al. [20] modeled how a four-segment PBN can follow an arbitrary 3D path by changing the relative offset between interlocked segments, and Secoli et al. [21] developed an adaptive closed-loop controller enabling PBNs to follow a prescribed curvilinear trajectory within tissue.

Planning an appropriate three-dimensional insertion path with the needle, which meets both kinematic constraints (such as the needle's maximum depth and minimum radius of curvature) and

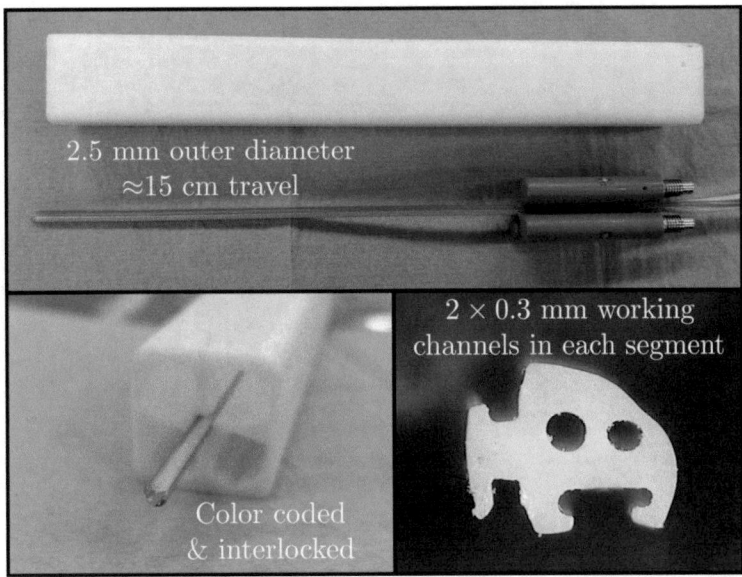

Fig. 3 The embodiment of the four-segment PBN extruded of a biocompatible material

avoids complex anatomical structures, would be taxing for the surgeon to perform manually. Therefore, an automatic path-planning algorithm is required for all needle steering systems, and for our PBN design. In our case [22], for instance, the algorithm can identify multiple viable paths and suggest a suitable entry area on the patient's skull (Fig. 4). The surgeon then chooses a path from several candidates that are generated by the path planner, which meets all necessary constraints and is subjectively preferred.

Because of the complexity of 3D path planning computations and the extent of data that has to be processed, the literature offers a variety of algorithms specifically designed for GPU and CPU acceleration, such as our own work on path planning [23] and real-time path replanning [22] to account for tissue motion and errors in path following.

3 Intraoperative Supervisory-Control

Intraoperatively, the use of supervisory-control robots in neurosurgery can be classified into two categories: passive and active systems [24]. In passive systems, the physical driving force to insert the surgical tool is provided by the surgeon, and the robot simply acts as an "intelligent guidance fixture." On the other hand, in active systems, a powered robot actively interacts with the patient [24]. Within the literature, the majority of supervisory-control robots are passive: the robot moves automatically into the desired

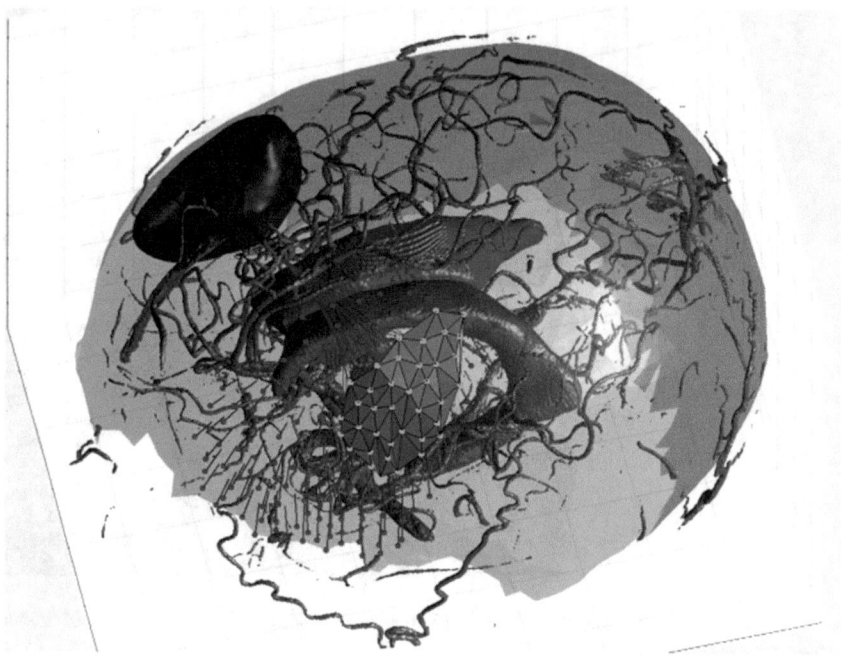

Fig. 4 Entry points and paths generated by the steerable needle path planner that can be chosen by the surgeon. (Reprinted from [22]. CC BY 4.0)

pose, then all of the joints are locked for safety (e.g., the Renaissance system).

Regardless of the type of interaction, registration is required to map the patient's location within the robot's coordinate system [9]. To compute the rigid body transformation from robot coordinates to "world coordinates," systems can use either framed or frameless approaches. In framed systems, a mechanical fixture is mounted onto the patient before preoperative image acquisition, and then a common reference frame is established by identifying features of the frame on the images, and securing the frame in a known configuration with respect to the robot. Conversely, in frameless systems, features are extracted directly from the anatomy of patients themselves in both the image- and robot-space. Examples include fiducial marker screws, skin adhesive markers, anatomical landmarks [25], and ultrasound registration [9]. Surface scanning of the patient's forehead and eyes can also be used to match the intraoperative robot position with preoperative images but with arguable accuracy and robustness [6, 26].

Following registration, the pose of the robot end effector can be computed, and the robot can move automatically as required. Passive robot systems provide guidance for the insertion of surgical instruments, such as biopsy needles, deep brain stimulation (DBS) electrodes, and pedicle screws, where constraints in at least five degrees of freedom [27] are necessary for appropriate guidance (i.e., a target point in space with two orientations, *see* Fig. 5).

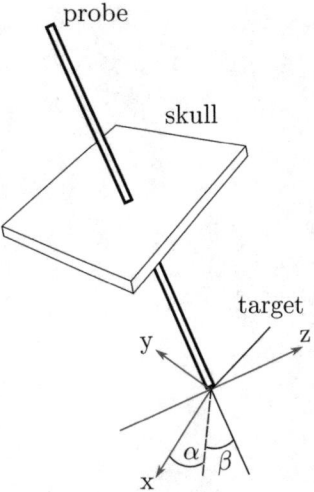

Fig. 5 The minimum five degree-of-freedoms to achieve a target in the brain. (Copyright ©1993 by Cambridge University Press. Reproduced with permission from [27])

An example of active robot system is the RobaCKa, which performs automatic milling of the skull according to a set of pre-determined trajectories [28]. The robot is equipped with force and torque sensors so that it can be stopped if the force and torque readings exceed a certain limit. In addition, an infrared navigation system also tracks the movement of the robot. However, due to accuracy limitations of the system in defining the boundary between the skull and dura mater, automatic milling is only performed for 50% of the skull's thickness. The remaining process is performed manually.

Another example of an active robot system was developed by Hu et al. [29]; it was designed to automatically remove tumor residues after conventional resection surgery. The tumor boundaries are detected using a fluorescent label and to remove the tumor, an ablation procedure was proposed which was modeled using a behavior tree framework.

Supervisory-control has also been implemented in both rigid needle or steerable needle systems. Amongst these, in addition to the minimum five degree-of-freedom requirement, Minerva [12] has an additional degree-of-freedom constraint to perform rigid needle insertion without having to change the position and orientation of the robot end effector. In the case of steerable needles, for instance with our PBN design, the needle delivery system can be mounted on a fine positioner, such as the Neuromate (Renishaw Inc., *see* Fig. 6), for accurate needle trocar placement, as was the case in the European Commission funded EDEN2020 project (www.eden2020.eu).

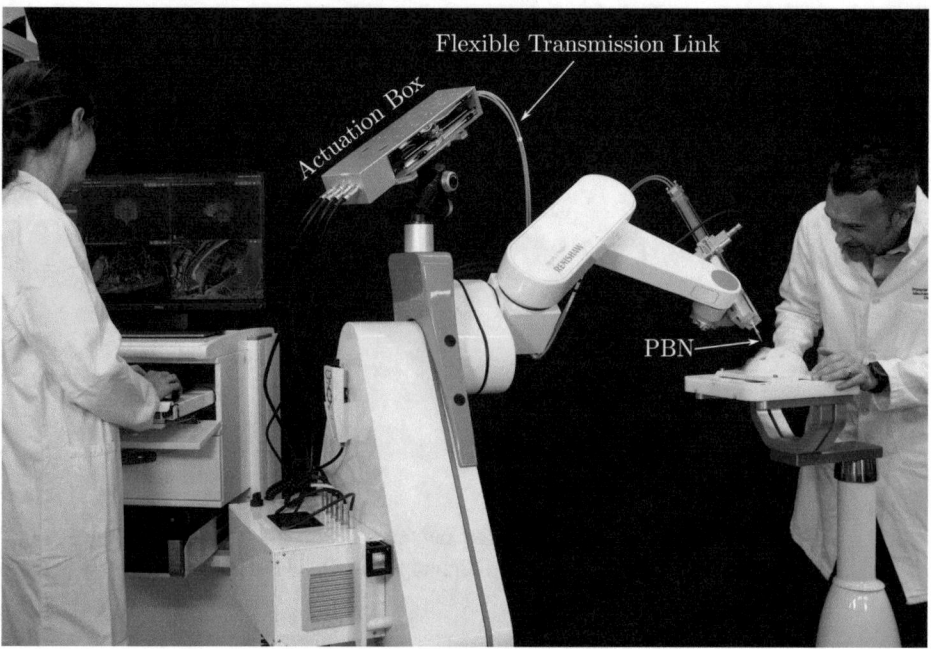

Flexible Transmission Link

Actuation Box

PBN

Fig. 6 The PBN system mounted on Renishaw's Neuromate, as part of the EDEN2020 project

The steerable needle should follow a certain path to achieve the target. In PBNs, the needle deflection is the result of complex tool–tissue interaction forces, which were explained via a mechanics-based steering model [20], which considers the multisegment structure as a set of multiple cantilevered Euler–Bernoulli beams.

To perform autonomous path following under the supervision of the surgeon, needle insertion can be performed through adaptive path-following control [30]. The controller consists of a high-level controller (HLC) and low-level controller (LLC). The HLC is inspired by path-following control for unmanned air vehicles. The HLC aims to generate the steering commands necessary to follow a desired path by dynamically altering the needle approach vector along the insertion path. The LLC is employed to follow these commands by creating optimal segment offsets based on the model developed in [20] and learns the nonlinear part of the model, adaptively, while the needle is inserted. A flowchart of the control strategy can be seen in Fig. 7, while Fig. 8 shows the trajectories achieved by the needle within a gelatin phantom.

As an alternative to supervisory-control, where needle insertion is entirely controlled by the robotic system, Matheson et al. [31] also proposed a surgeon-in-the-loop control strategy, where path following is performed cooperatively. Cartesian commands directing the needle tip along a continuously adapting 3D path are imparted by the surgeon via a haptic joystick (Fig. 9), then curvature-offset mapping is performed autonomously by the

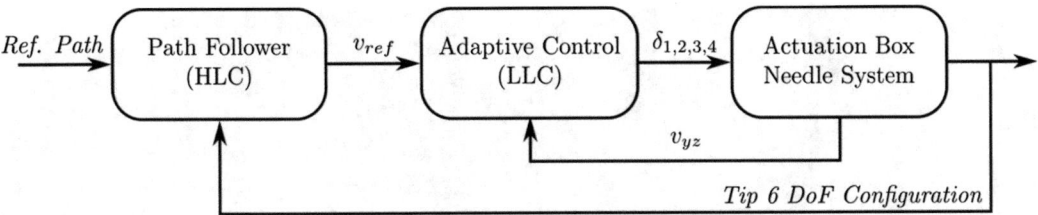

Fig. 7 Flowchart of the steerable needle control strategy with inner loop control to track the command (v_{ref}) from the HLC. The LLC gives relative offset configuration command ($\delta_{1,2,3,4}$) to the actuation box needle system. The HLC is in charged to minimize the distance between the reference path with the needle tip. (Copyright © 2016 by IEEE. Reprinted with permission from [30])

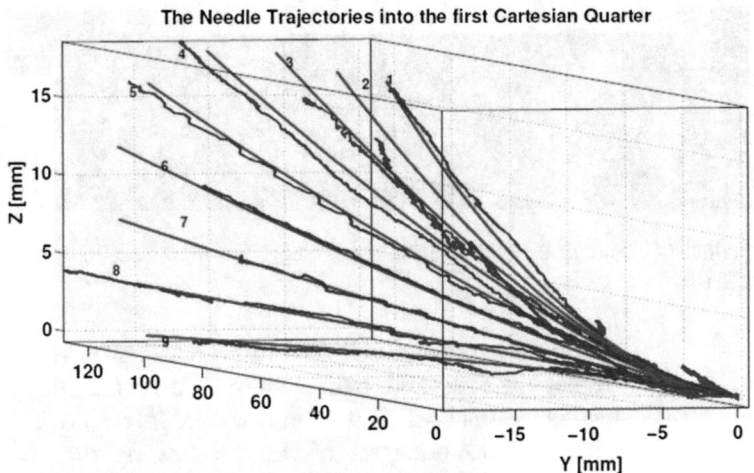

Fig. 8 Trajectories of the PBN tracked using electromagnetic tracker. Red curves represent reference paths, and blue curves represent the needle's average path. (Copyright © 2018 by IEEE. Reprinted with permission from [21])

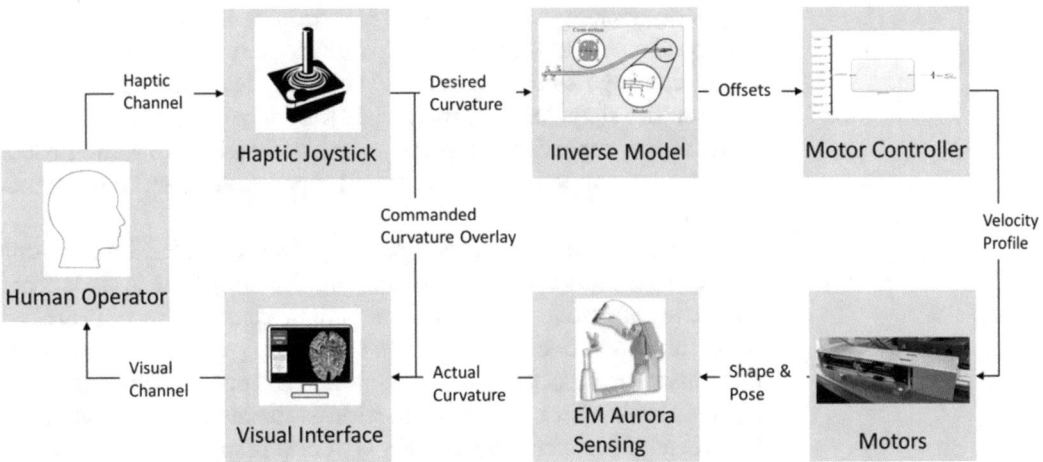

Fig. 9 System architecture for a surgeon-in-the-loop control strategy. (Copyright © 2019 by IEEE. Reprinted with permission from [31])

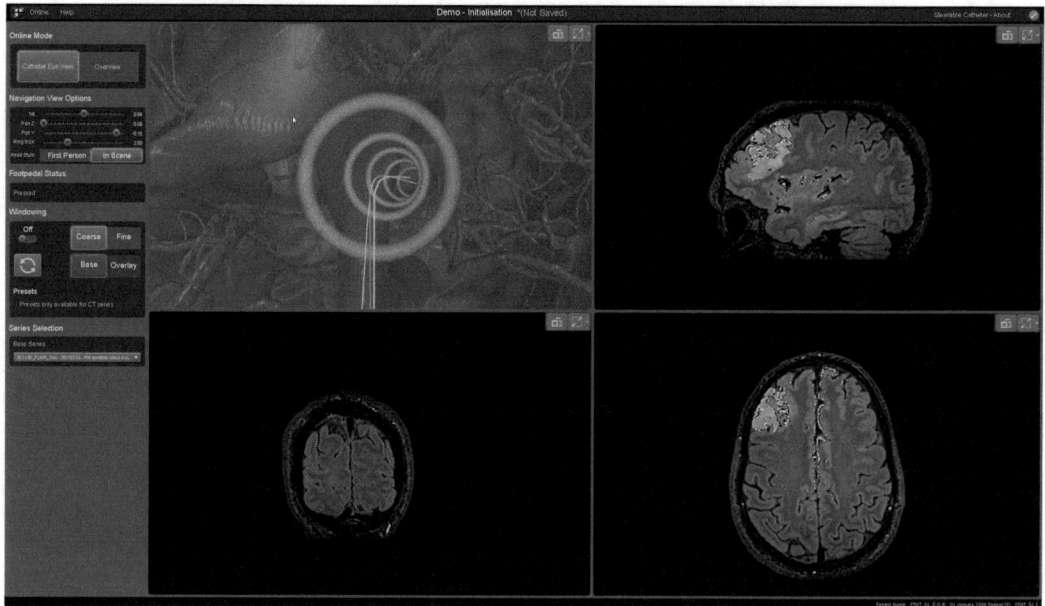

Fig. 10 Visual interface of the surgeon-in-the-loop control strategy. (Copyright © 2019 by IEEE. Reprinted with permission from [31])

system to execute the desired motion. A visual front end (Fig. 10) displays the desired path (white), current heading (blue), and instantaneous Cartesian reference pose (green) to aid the surgeon, with a color-coded set of rings indicating the current tracking accuracy. In this servo-assisted embodiment, the surgeon and the robotic system make use of environmental sensors to reach the desired target pose collaboratively (electromagnetic sensor in Fig. 9).

To track the needle inside tissue, Secoli et al. [30] used electromagnetic tracking sensors (Aurora, Northern Digital Inc.) embedded in each needle segment. More recently, a system based on Fiber Bragg Gratings is being developed to track the needle tip, as well as to sense the shape each segment [32]. By using shape sensing, the surgeon can supervise the shape of the catheter so that the procedure can be stopped in case imminent separation between segments is detected.

As any form of neurosurgery involves complex interactions between surgical instruments and highly compliant and delicate tissue, preoperative images of the patient cannot always be relied upon intraoperatively. During the supervisory-control insertion of rigid instruments, however, tissue deformations are often ignored because their effect on target motion is deemed to be negligible. Conversely, for steerable needle systems, tissue deformation can have significant consequences for the system's ability to follow a prescribed curvilinear path to reach a deep-seated tissue target.

Consequently, a number of intraoperative imaging modalities, such as interventional MRI (MRI Interventions, Inc.), O-arm imaging [33], and C-arm imaging [34], can be employed. Additionally, intraoperative ultrasound has recently been proposed by Riva et al. [35] to guide the insertion process on the basis of updated images, where preoperative MRI is fused with intraoperative ultrasound to generate highly resolved temporal and spatial volumes for guidance. Finally, since steerable needles can follow curvilinear paths to avoid obstacles, some researchers have been exploring the use of environmental sensors to detect specific anatomical features, such as vessels and tissue boundaries, "on-the-fly" during the insertion process. In our most recent work, forward-looking sensors, coupled with a deep learning approach, have been embedded within a miniature PBN prototype to successfully investigate the presence of vessels in front of the tip of the needle [36].

4 Conclusion

The majority of neurosurgical robots to date are based on the supervisory-control method. To perform supervisory-control, the robot should have the ability to perform the task automatically under surgeon supervision. During the preoperative procedure, robot systems can provide suggestions to the surgeon during the surgical planning process. Intraoperatively, after the surgeon has confirmed the procedure that will be executed, a passive supervisory-control robot will move to the desired position and orientation, before the surgeon performs instrument insertion manually. Conversely, in active robotic systems, the procedure is performed entirely by the robot under the supervision of the surgeon so that there is direct interaction between the robot and the patient.

With active systems, appropriate modeling of the interaction between the robot and the tissue is required to perform a safe and an accurate procedure. Additionally, the surgeon needs continuous feedback about the procedure that is performed by the robot so that, in case of an emergency, the procedure can be interrupted. Consequently, intraoperative imaging approaches to monitor the surgical area, and embedded sensing systems for accurate, real-time in situ measurements, have also been proposed to obtain up-to-date information of the robot state and the patient.

References

1. Sheridan TB (2012) Human supervisory control. In: Salvendy G (ed) Handbook of human factors and ergonomics, 4th edn. Wiley, New York, pp 990–1015
2. Kwoh YS, Hou J, Jonckheere EA, Hayati S (1988) A robot with improved absolute positioning accuracy for CT guided stereotactic brain surgery. IEEE Trans Biomed Eng 35:153–160
3. Harris SJ, Arambula-Cosio F, Mei Q et al (1997) The Probot—an active robot for prostate resection. Proc Inst Mech Eng H J Eng Med 211:317–325. https://doi.org/10.1243/0954411971534449
4. Moustris GP, Hiridis SC, Deliparaschos KM, Konstantinidis KM (2011) Evolution of autonomous and semi-autonomous robotic surgical systems: a review of the literature. Int J Med Robot Comput Aided Surg 7:375–392. https://doi.org/10.1002/rcs
5. Sheridan T (1984) Supervisory control of remote manipulators, vehicles and dynamic process: experiments in command and display aiding, pp 49–137
6. Joskowicz L, Shamir R, Freiman M et al (2006) Image-guided system with miniature robot for precise positioning and targeting in keyhole neurosurgery. Comput Aided Surg 11:181–193. https://doi.org/10.1080/10929080600909351
7. Marcus HJ, Seneci CA, Payne CJ et al (2014) Robotics in keyhole transcranial endoscope-assisted microsurgery: a critical review of existing systems and proposed specifications for new robotic platforms. Neurosurgery 10:84–95. https://doi.org/10.1227/NEU.0000000000000123
8. Marcus HJ, Vakharia VN, Ourselin S et al (2018) Robot-assisted stereotactic brain biopsy: systematic review and bibliometric analysis. Childs Nerv Syst 34:1299–1309. https://doi.org/10.1007/s00381-018-3821-y
9. Li QH, Zamorano L, Pandya A et al (2002) The application accuracy of the NeuroMate robot—a quantitative comparison with frameless and frame-based surgical localization systems. Comput Aided Surg 7:90–98. https://doi.org/10.1002/igs.10035
10. Lefranc M, Peltier J (2016) Evaluation of the ROSA™ Spine robot for minimally invasive surgical procedures. Expert Rev Med Devices 13:899–906. https://doi.org/10.1080/17434440.2016.1236680
11. Glauser D, Fankhauser H, Epitaux M et al (1995) Neurosurgical robot Minerva: first results and current developments. Comput Aided Surg 1:266–272. https://doi.org/10.3109/10929089509106332
12. Hefti JL, Epitaux M, Glauser D, Fankhauser H (1998) Robotic three-dimensional positioning of a stimulation electrode in the brain. Comput Aided Surg 3:1–10. https://doi.org/10.1002/(SICI)1097-0150(1998)3:1<1::AID-IGS1>3.0.CO;2-3
13. Grimm F, Naros G, Gutenberg A et al (2015) Blurring the boundaries between frame-based and frameless stereotaxy: feasibility study for brain biopsies performed with the use of a head-mounted robot. J Neurosurg 123:737–742. https://doi.org/10.3171/2014.12.JNS141781
14. Lévesque MF, Parker F (1999) MKM-guided resection of diffuse brainstem neoplasms. Stereotact Funct Neurosurg 73:15–18. https://doi.org/10.1159/000029744
15. Willems PWA, Noordmans HJ, van der Sprenkel JWB et al (2001) An MKM-mounted instrument holder for frameless point-stereotactic procedures: a phantom-based accuracy evaluation. J Neurosurg 95:1067–1074. https://doi.org/10.3171/jns.2001.95.6.1067
16. Secoli R, Matheson E, Rodriguez y Baena F (2018) A modular robotic catheter driver for programmable bevel-tip steerable needles. In: Proceedings - Hamlyn Symposium on Medical Robotics 2018
17. Chan F, Kassim I, Lo C et al (2009) Image-guided robotic neurosurgery-an in vitro and in vivo point accuracy evaluation experimental study. Surg Neurol 71:640–647. https://doi.org/10.1016/j.surneu.2008.06.008
18. Van De Berg NJ, Van Gerwen DJ, Dankelman J, Van Den Dobbelsteen JJ (2015) Design choices in needle steering - a review. IEEE/ASME Trans Mechatron 20:2172–2183. https://doi.org/10.1109/TMECH.2014.2365999
19. Ko SY, Davies BL, Rodriguez y Baena F (2010) Two-dimensional needle steering with a "programmable bevel" inspired by nature: modeling preliminaries. In: IEEE/RSJ 2010 International Conference on Intelligent robots and systems, IROS 2010 - Conference Proceedings, pp 2319–2324
20. Watts T, Secoli R, Rodriguez y Baena F (2019) A mechanics-based model for 3-D steering of programmable bevel-tip needles. IEEE Trans

Robot 35:371–386. https://doi.org/10.1109/TRO.2018.2879584

21. Secoli R, Rodriguez y Baena F (2018) Experimental validation of curvature tracking with a programmable bevel-tip steerable needle. In: 2018 International Symposium on Medical robotics, ISMR 2018, pp 1–6

22. Pinzi M, Galvan S, Rodriguez y Baena F (2019) The adaptive hermite fractal tree (AHFT): a novel surgical 3D path planning approach with curvature and heading constraints. Int J Comput Assist Radiol Surg 14:659–670. https://doi.org/10.1007/s11548-019-01923-3

23. Liu F, Garriga-Casanovas A, Secoli R, Rodriguez y Baena F (2016) Fast and adaptive fractal tree-based path planning for programmable bevel tip steerable needles. IEEE Robot Autom Lett 1:601–608. https://doi.org/10.1109/LRA.2016.2528292

24. Nathoo N, Çavuşoğlu MC, Vogelbaum MA, Barnett GH (2005) In touch with robotics: neurosurgery for the future. Neurosurgery 56:421–431. https://doi.org/10.1227/01.NEU.0000153929.68024.CF

25. Ortmaier T, Weiss H, Döbele S, Schreiber U (2006) Experiments on robot-assisted navigated drilling and milling of bones for pedicle screw placement. Int J Med Robot Comput Aided Surg 2:350–363. https://doi.org/10.1002/rcs

26. Zimmermann M, Krishnan R, Raabe A, Seifert V (2004) Robot-assisted navigated endoscopic ventriculostomy: implementation of a new technology and first clinical results. Acta Neurochir 146:697–704. https://doi.org/10.1007/s00701-004-0267-7

27. Glauser D, Flury P, Burckhardt CW (1993) Mechanical concept of the neurosurgical robot "Minerva". Robotica 11:567–575

28. Korb W, Engel D, Boesecke R et al (2003) Development and first patient trial of a surgical robot for complex trajectory milling. Comput Aided Surg 8:247–256. https://doi.org/10.3109/10929080309146060

29. Hu D, Gong Y, Seibel EJ et al (2018) Semi-autonomous image-guided brain tumour resection using an integrated robotic system: a bench-top study. Int J Med Robot Comput Assist Surg 14:1–15. https://doi.org/10.1002/rcs.1872

30. Secoli R, Rodriguez y Baena F (2016) Adaptive path-following control for bio-inspired steerable needles. In: Proceedings of the IEEE RAS and EMBS International Conference on Biomedical robotics and biomechatronics, pp 87–93

31. Matheson E, Secoli R, Galvan S, Rodriguez y Baena F (2019) Human-robot visual interface for 3D steering of a flexible, bioinspired needle for neurosurgery. In: 2019 IEEE/RSJ International Conference on Intelligent robots and systems

32. Khan F, Denasi A, Barrera D et al (2019) Multi-core optical fibers with Bragg gratings as shape sensor for flexible medical instruments. IEEE Sensors J 19:5878–5884. https://doi.org/10.1109/JSEN.2019.2905010

33. Smith AP, Bakay RAE (2011) Frameless deep brain stimulation using intraoperative O-arm technology: clinical article. J Neurosurg 115:301–309. https://doi.org/10.3171/2011.3.JNS101642

34. Mirzayan MJ, Von Roden M, Bulacio J et al (2016) The usefulness of intraoperative cerebral C-arm CT angiogram for implantation of intracranial depth electrodes in stereotactic electroencephalography procedure. Stereotact Funct Neurosurg 94:10–17. https://doi.org/10.1159/000431372

35. Riva M, Hennersperger C, Milletari F et al (2017) 3D intra-operative ultrasound and MR image guidance: pursuing an ultrasound-based management of brainshift to enhance neuronavigation. Int J Comput Assist Radiol Surg 12:1711–1725. https://doi.org/10.1007/s11548-017-1578-5

36. Virdyawan V, Rodriguez y Baena F (2019) A long short-term memory network for vessel reconstruction based on laser Doppler flowmetry via a steerable needle. IEEE Sensors J. https://doi.org/10.1109/JSEN.2019.2934013

Chapter 3

A Teleoperated Surgical Robot System

Andria A. Remirez, Margaret F. Rox, Trevor L. Bruns, Paul T. Russell, and Robert J. Webster III

Abstract

This chapter reviews a teleoperated surgical robotic system that we have developed over the past several years at Vanderbilt University. It delivers needle-sized instruments into the human body that are able to move in a tentacle-like manner in the sense that they can controllably bend and elongate. Preclinical studies on this class of robots (by both our group and others) have investigated the feasibility of using them for intracerebral hemorrhage aspiration, thermal ablation to treat epilepsy, endoscopic third ventriculostomy, endoscopic colloid cyst removal, and endonasal pituitary surgery. This chapter initially describes the system from the perspective of endonasal pituitary surgery, but also includes a section at the end summarizing how the same basic robot concept can be applied in the other neurosurgical contexts mentioned above. We believe that one day, a system of the type described in this chapter will provide a "da Vinci-like" platform where the surgeon teleoperates the robot from a control console, and the robot makes minimally invasive procedures much easier for the surgeon.

Key words Surgical robots, Teleoperated robots, Flexible robots, Continuum robots, Concentric tube robots, Steerable needles, Endoscopic surgery

---

1 Introduction

Surgical robots today can be broadly categorized into three basic types, according to the way they interact with the physician [1]. There are robots that perform the surgeon's pre-operative plan autonomously and accurately (e.g. the TSolution One system, formerly marketed as RoboDoc [2]), robots that are remotely controlled (i.e. "teleoperated") by the physician with no autonomy (e.g. the da Vinci Surgical system [3]), and robots where the human and robot cooperatively hold the surgical instrument (e.g. the Mako system [4]). Teleoperated robots can provide dexterity and precision in minimally invasive surgery while keeping the surgeon in direct control of his or her tools.

This chapter will focus primarily on a new kind of teleoperated surgical robot that is particularly advantageous for neurosurgery. It

Hani J. Marcus and Christopher J. Payne (eds.), *Neurosurgical Robotics*, Neuromethods, vol. 162, https://doi.org/10.1007/978-1-0716-0993-4_3, © Springer Science+Business Media, LLC, part of Springer Nature 2021

has manipulators that are needle-sized, are able to bend and elongate, and can be directly controlled by the surgeon through a user interface. Robots that can bend and elongate are known in general as "continuum robots" because they are continuously flexible [5]. Because of their ability to take on a variety of curved shapes, they can maneuver around anatomical obstacles and travel through curvilinear paths, making them well-suited to a variety of medical procedures. Indeed, familiar interventional instruments such as flexible endoscopes and catheters are (non-robotic) continuum devices which have been used in medicine for many years. Continuum robots can be constructed in a variety of ways [5], including using pull-wires [6–10], elastic rods [11, 12], or pressurized fluid-filled chambers [13–15].

In this chapter, we focus on one particular type of continuum robot known as a concentric tube robot [16] that is uniquely well-suited for a variety of neurosurgical procedures in which dexterity is extremely valuable and small diameter is essential. A concentric tube robot (see Fig. 1) consists of several flexible tubes (typically made from superelastic nitinol), which are pre-curved at their tips and then nested inside one another. The tubes are then grasped at their proximal ends and small electric motors are used to axially rotate and telescopically extend and retract each tube with respect to the others. Telescoping motions produce elongation in the collection of tubes, while axial rotation of each tube causes the collection of tubes to bend into curved configurations, as the pre-curved tube tips cause each other to bend and twist. Through the use of a mechanics-based model [16, 17], the bending and twisting can be described mathematically, enabling a computer to determine how much to rotate and extend each tube in order to move the tip of the device in the surgeon's desired direction.

Researchers have proposed concentric tube robots for a variety of clinical applications, including prostate surgery [18], lung biopsy [19], and cardiac interventions [20]. In the realm of neurosurgery, concentric tube robots have been used for pituitary tumor removal [21], MRI-guided ablation for epilepsy treatment [22], evacuation of intracerebral hemorrhages [23], and enhancing endoscopic

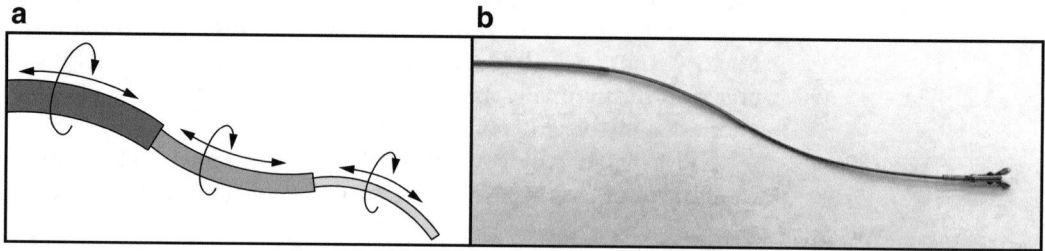

Fig. 1 Concentric tube robots, consisting of nested, flexible, pre-curved tubes. (**a**) Illustration of the motions which can be applied to each tube. (**b**) A prototype concentric tube robot with a gripper at the tip

access to the ventricles [24–26]. In this chapter, we begin by providing an overview of the development of a robotic system capable of deploying several concentric tube manipulators and an endoscope through the nostrils, providing dexterous teleoperated tools for procedures in the sinuses and at the skull base. At the end of the chapter, we discuss other neurosurgical applications of concentric tube robots.

2 System Overview

Our system consists of two basic components: the robot and the surgeon interface console, as shown in Fig. 2. Mounted over the patient, the robot deploys up to three steerable instruments into the nostrils, along with a variable view angle endoscope. The physician controlling these instruments is seated at the surgeon interface console, where a large monitor displays the endoscopic view and input devices allow him or her to control the needle-sized instruments.

The robot is mounted on a lockable, counter-balanced arm so that it can be easily positioned over the patient as desired and locked in place during the surgery. Up to three different instruments extend from the front end of the robot through an instrument channel, as shown in Fig. 3, which can be inserted into the patient's nostril to a desired depth. An adjustable view angle rigid

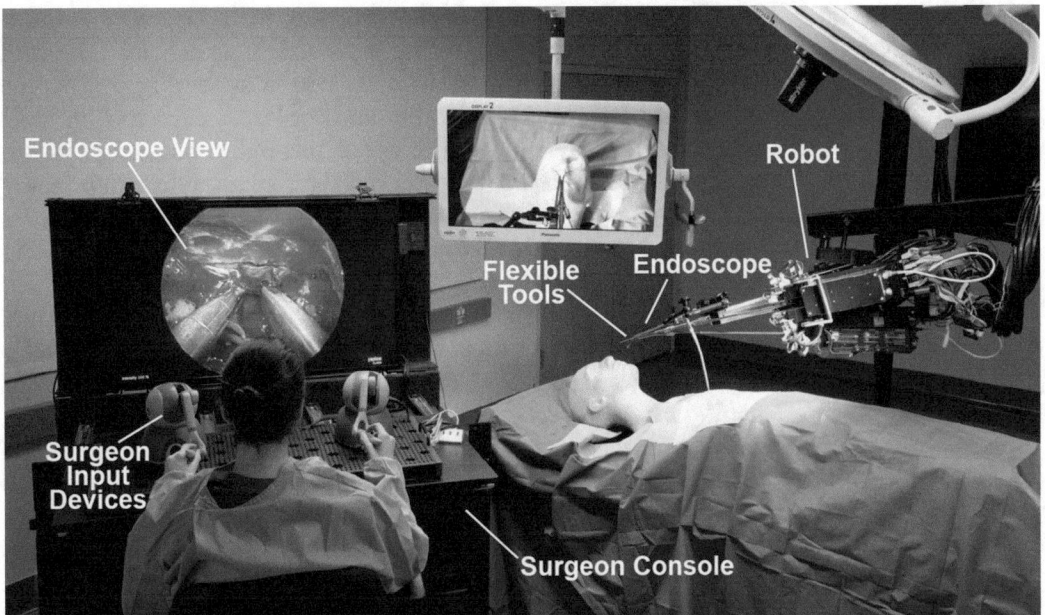

Fig. 2 Complete endonasal surgical system, including the robot mounted above the patient and the surgeon console

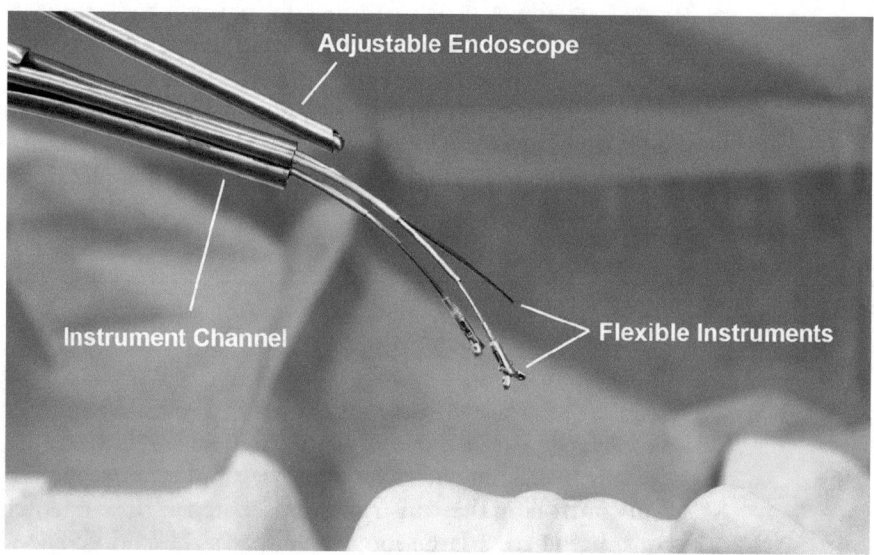

Fig. 3 Instrument channel which aligns the tools at the front end of the robot, with variable view angle endoscope mounted on a lockable positioning arm

endoscope, the EndoCAMeleon (Karl Storz SE & Co. KG), is used for visualization. Its view angle can be adjusted within a 75° range without moving the endoscope itself. The endoscope is 4 mm in diameter and is mounted on a lockable arm at the front of the robot (see Fig. 3), enabling the user to position it as desired, lock it in place, and reposition it as needed.

The robotic system is highly modular, with the ability to interchange tools prior to or during a procedure as needed. The tool modules, shown attached to the base unit in Fig. 4a, come equipped with the desired tubes and tool tips. Each tool module mates to the base unit of the robot through a coupling system, which connects it to the motors, and the tool modules can be easily added to or removed from the base unit via a quick-latch feature, as shown in Fig. 4b. This design scheme not only enables quick and easy tool changes, but also facilitates the use of a sterility method inspired by the da Vinci system. Sterility can pose a major challenge in surgical robotics, as most motors and electronics cannot be easily sterilized. As such, these elements are isolated from the patient behind a sterile drape, with only sterile components (tool modules, endoscope, instrument channel, etc.) remaining on the patient-side of the drape, as shown in Fig. 4c. To accomplish this, the drape must be produced with custom adapters built into it, through which the spring-loaded couplings can join the base unit and the tool modules.

Tool cartridges can be equipped with a variety of different tool tips. Each cartridge is equipped with either a mechanism to pull a wire (which is useful for tasks like operating grippers) or a

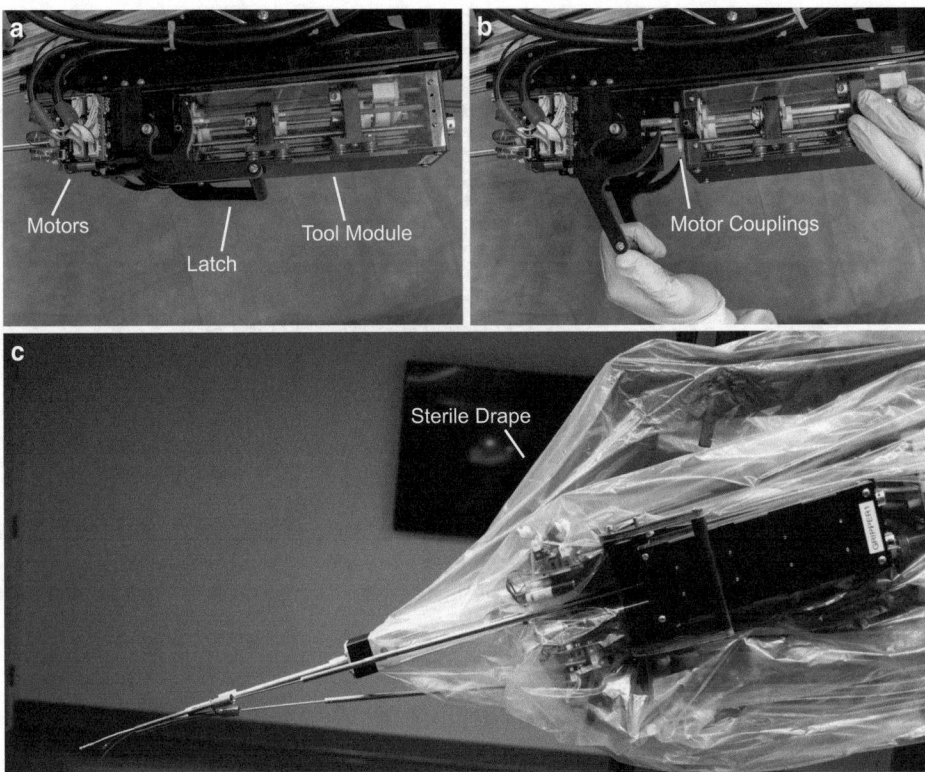

Fig. 4 (**a**) Tool module attached to the base unit of the robot. (**b**) Tool module being detached from the system using the quick-latch feature. (**c**) Sterile drape covering the non-sterile components of the system, while sterile components remain on the outside of the drape

mechanism to rotate an inner wire or tube (which is useful for providing direct axial rotation of a tool such as a curette). Figure 5a shows several example tool tips which are useful for many transnasal procedures, including a gripper and curette. In addition, suction or irrigation can be provided directly through the innermost tube of a concentric tube robot arm.

To enhance articulation at the tool tips and increase the range of achievable orientations, a miniature "wrist," as shown in Fig. 5b, can be used [27]. These wrists can be built directly into the innermost tube of a concentric tube robot arm by cutting away sections of material to produce a direction of preferential bending. Articulation can then be achieved by using a motor to pull on a wire affixed to the tip of the tube.

A custom surgeon console (see Fig. 2) provides the interface between the physician and the robot. It provides multiple input devices to enable control of more than one tool simultaneously, along with foot pedals which can be used to control which tools are active. Endoscope camera footage is provided on a large display built into the surgeon console. The entire console is designed as a

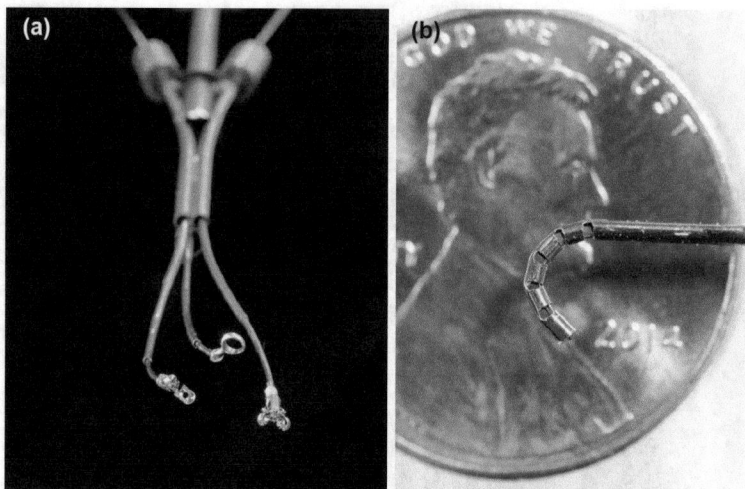

Fig. 5 (a) Examples of the types of tool tips which can be attached to the flexible instruments, including grippers and a curette. (b) High curvature "wrist" prototype which can be used to augment angulation at the tip of the flexible instruments

wheeled, mobile cart, facilitating easy transport within a hospital or surgery center.

The robotic control software is based on the scheme described by Burgner et al. [28], in which the velocity of the surgeon's input motion dictates the desired velocity of the instrument tip. In addition to adding dexterity and providing an intuitive mapping between the motion of input devices and robotic tools, the teleoperation framework can enable several potentially useful features such as motion scaling, tremor cancellation, or virtual fixtures to limit motions relative to certain anatomical features [29].

3 Clinical Applications

3.1 Pituitary Tumor Resection

Several experimental studies have been conducted to validate the use of concentric tube robots for endonasal surgery. Experiments performed by Swaney et al. in [21] verified that a pituitary adenoma can be removed using concentric tube manipulators. A phantom tumor was constructed using ballistics gel to closely simulate the consistency of a pituitary adenoma, and placed inside an anatomical skull model, in which a sphenoidotomy had been performed to provide transnasal access to the site of the tumor. Under endoscopic visualization as shown in Fig. 6, a surgeon used the robot to remove as much of the phantom tumor as possible using a curette and suction through the tubes. In 20 trials, the average percentage of tumor removed was 79.8±5.9%, which is considered clinically

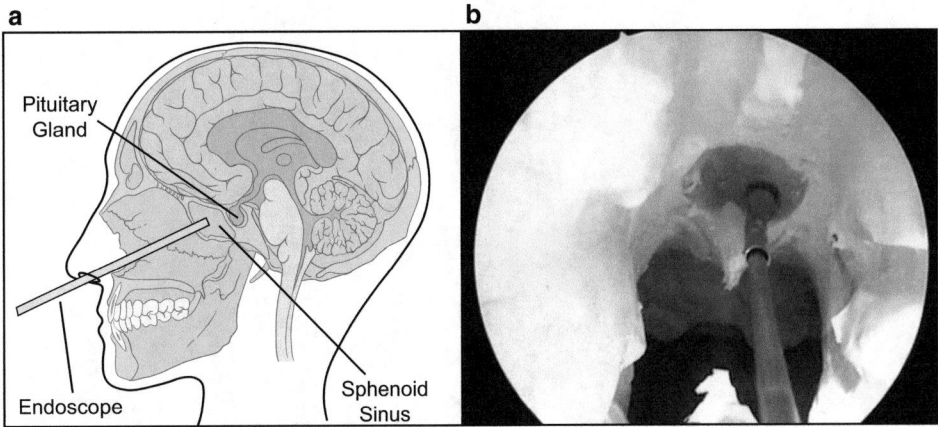

Fig. 6 (**a**) Illustration of the transnasal approach to pituitary tumor removal. (**b**) Endoscopic view of the surgical site during phantom pituitary tumor resection experiments

useful for the benign and slow-growing pituitary adenomas targeted.

Further work by Wirz et al. [30] used the same method for pituitary tumor resection in phantom models to demonstrate the viability of telerobotic surgery over long distances with a concentric tube robot system. During these experiments, one trial involved having the surgeon in the same room as the robot, at Vanderbilt University (Nashville, Tennessee, USA). In a second trial, the robot was at the University of North Carolina at Chapel Hill (Chapel Hill, North Carolina, USA), while the surgeon was at Vanderbilt, approximately 800 km away from the robot. In both cases, the surgeon was able to successfully resect the tumor, and any latency due to the Internet connection in the long-distance surgery was imperceptible to the physician.

3.2 Endoscopic Intraventricular Neurosurgery

Concentric tube robots can also be used for endoscopic neurosurgery, as described by Hendrick et al. [26]. Rigid endoscopy in the brain is much less invasive than the open approach, but typically necessitates tilting within the brain tissue, which can compress and damage critical structures. Concentric tube robots eliminate the need for this tilting, since they can move independently at the surgical site without requiring any motion of the endoscope itself. Using a robot (shown in Fig. 7) that coupled to a standard clinical rigid endoscope, Hendrick compared a manual neuroendoscopic procedure to a robotic approach [26]. The example of colloid cyst removal was used in a set of experiments. A phantom brain made of silicone was constructed based on a CT scan, and a colloid cyst phantom made from dilute agarose gel and stretched Parafilm was placed within the ventricles.

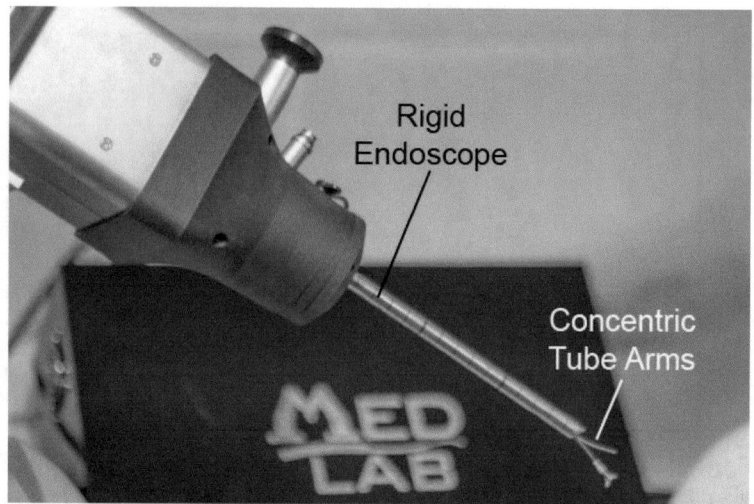

Fig. 7 Neuroendoscopic robot, with two concentric tube manipulators deployed through a standard rigid endoscope

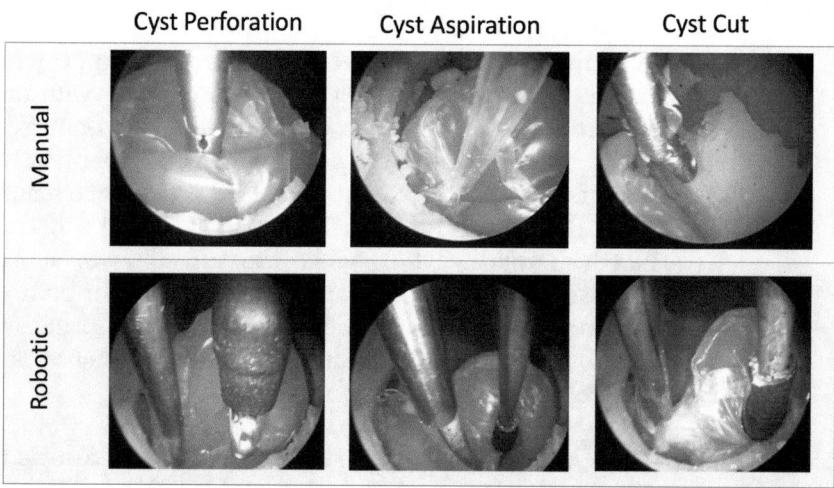

Fig. 8 The procedure was divided into three stages: cyst perforation, aspiration, and cut. The manual procedure required a complete tool change between stages, and the camera field of view moved with the tools. During the robotic procedure, the surgeon's field of view was stable, and he could use two arms to complete tasks instead of one, with no tool changes

Two neurosurgeons performed the manual procedure, with an attending surgeon holding the endoscope and a surgical fellow operating the tools, demonstrating that maximum endoscope tilt for the procedure was 17.1°. A single neurosurgeon completed the robotic procedure, with a maximum of 1.2° of endoscope tilt observed, as the robot was mounted on a stationary arm, while the surgeon teleoperated the instruments from a nearby console. A comparison of different stages of the procedure with the manual approach vs. the robotic approach can be seen in Fig. 8, indicating

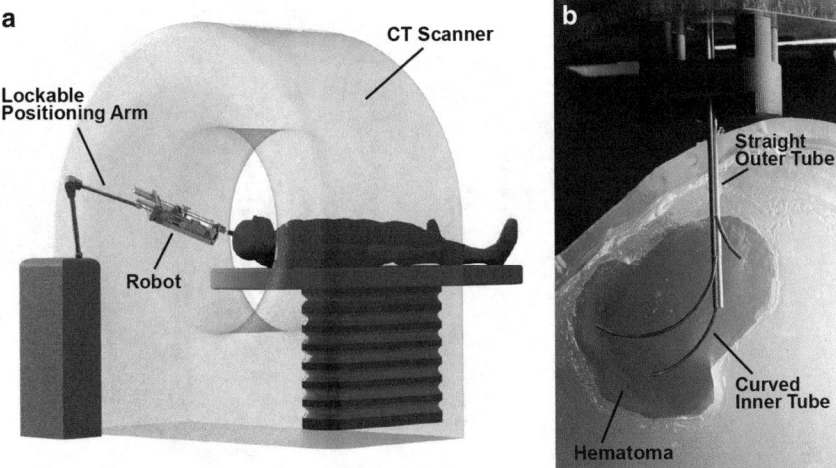

Fig. 9 (**a**) Conceptual illustration of an in-scanner robotic approach to intracerebral hemorrhage evacuation. (**b**) Two-tube instrument inside of a simulated hematoma, with multiple positions of the instrument superimposed onto each other to demonstrate its range of motion

that the robotic instruments enabled the surgeon to perform certain tasks within the procedure using different techniques than he could with the manual approach. In addition to dramatically reducing endoscope tilt, the robotic approach reduced the number of surgeons performing the surgery from two to one.

3.3 Intracerebral Hemorrhage Evacuation

Burgner et al. presented a concentric tube robot system for treating intracerebral hemorrhage (in which blood pools in the brain, compressing and damaging the surrounding tissue) [23]. While it is generally believed that decompression by removal of hematoma material can save surrounding brain tissue, the surgery itself is invasive, with a substantial risk of complications using current tools. To reduce the invasiveness of the procedure, the robotic system is used to deploy a small diameter, straight tube through a burr hole and toward the center of the hematoma. A smaller, flexible, curved tube is attached to a vacuum pump and deployed through the straight outer tube. Rotation of the inner tube and telescopic extension of both tubes can be used to debulk the hematoma, as illustrated in Fig. 9b. Godage et al. demonstrated in phantom models that performing the procedure in a CT scanner with intraoperative imaging, as in Fig. 9a, enables the user to replan the motion of the tubes as the volume of the hematoma changes during decompression [31]. This has the effect of reducing the risk to surrounding brain tissue and enabling more hematoma material to be removed safely.

3.4 MRI-Guided Ablation for Epilepsy Treatment

Up until now we have described concentric tube robots as small, flexible manipulators. However it is also possible to use them as steerable needles in a tissue-embedded context. Here, they are used to target specific anatomical sites while traveling through soft

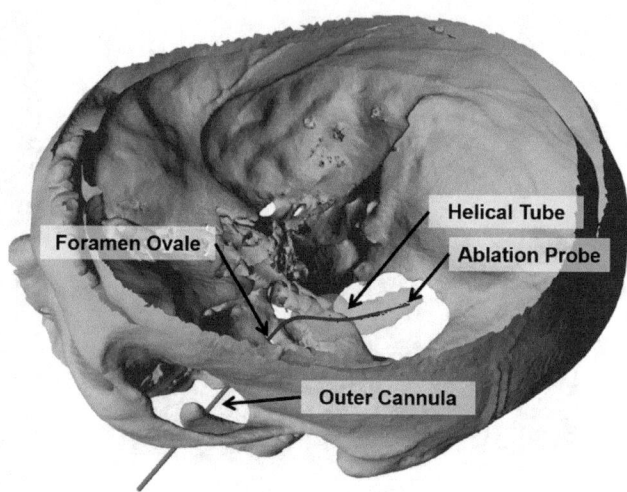

Fig. 10 Illustration of the deployment of a helically curved tube for transforamenal ablation of the hippocampus

tissue, open spaces, or narrow lumens in the anatomy. Care must be taken in both tube shaping and actuation in order to make a concentric tube robot behave in this way, but it is possible [32].

Targeted ablation for the treatment of epilepsy is one application where a curved needle of this type could be useful. Surgical resection or thermal ablation of the amygdala and hippocampus has been shown to reduce seizures in many patients whose epilepsy does not respond well to drugs [33]. Comber et al. proposed the use of concentric tube robots to access this region using a transforamenal approach [34], which eliminates the need to drill through bone, as illustrated in Fig. 10. In this approach, the tubes closely follow the midline of the hippocampus to achieve the best possible distribution of thermal energy. The concept is to use intraoperative MRI guidance and MR thermometry during this procedure, so a robotic system using custom-designed pneumatic actuators (as opposed to the more commonly used electric motors) was designed for in-MRI concentric tube insertion [35].

4   Conclusion

Concentric tube robots have already shown great promise for augmenting a surgeon's abilities when operating through narrow passageways. In endonasal surgery, these robots have the potential to improve dexterity and to navigate around corners to access difficult-to-reach regions of the skull base and sinuses. These advantages could ultimately make transnasal procedures easier for surgeons and even increase the number of surgeries which can be done using this minimally invasive approach. Concentric tube robots have also proven useful as tools to be deployed through rigid endoscopes, enabling, for example, intraventricular procedures that do not require tilting the endoscope to achieve dexterous tissue manipulation. In addition, they can be used as steerable suction devices for minimally invasive debulking of intracerebral hemorrhages and steerable needles to deliver ablative treatments into the brains of epilepsy patients.

As researchers continue to make advancements in the science of concentric tube robots, the clinical benefits associated with these robots could become even greater. For example, research into the unique ability of continuum robots to act as "intrinsic force sensors" [36] can enable haptic feedback, in which motors in the input device can be used to restore the surgeon's sensation of applied forces at the surgical site, which today is typically sacrificed when using teleoperated surgical robots [37]. Force sensing could also enable more advanced control techniques that further augment the surgeon's capabilities, such as limiting the amount of force a tool will apply to a particular anatomical structure. Investigation into optimal placement of sensors in continuum robots is promising for further improving accuracy [38, 39], and development and evaluation of user interfaces for continuum robots may enable increasingly intuitive control of these devices [40, 41]. In addition, design optimization of continuum robots for specific surgical sites and tasks can be used to improve dexterity, workspace, and other desirable traits within the context of individual procedures [20, 42–47]. Thus, a wide variety of possible features and improvements are actively being researched by roboticists.

Rigorously evaluating concentric tube robot systems for various neurosurgical procedures will also require further experimental studies as these robots move from a laboratory setting toward clinical adoption. Such tests are likely to include more direct experimental comparisons against manual approaches, as well as cadaver and animal studies, followed eventually by clinical studies in patients. The various systems and applications we have discussed in this chapter currently exist at different points along this spectrum of development and evaluation, but none has yet moved beyond animal and cadaver testing. Still, the vast majority of the design and control challenges for these robots have been solved, and it is likely that we will see them in the operating room in human use within the next several years. We anticipate that as these systems move toward clinical use, even more procedures will be identified which can benefit from the use of these small, flexible robotic tools.

References

1. Taylor RH, Menciassi A, Fichtinger G, Fiorini P, Dario P (2016) Medical robotics and computer-integrated surgery. Springer International Publishing, Cham, pp 1657–1684

2. Bargar WL, Bauer A, Börner M (1998) Primary and revision total hip replacement using the Robodoc (R) System. Clin Orthop Relat Res 354:82–91

3. DiMaio S, Hanuschik M, and Kreaden U (2011) The da Vinci surgical system. In Surgical Robotics. Springer, Boston, MA, pp 199–217

4. Pearle AD, O'Loughlin PF, Kendoff DO (2010) Robot-assisted unicompartmental knee arthroplasty. J Arthroplast 25 (2):230–237

5. Burgner-Kahrs J, Rucker DC, Choset H (2015) Continuum robots for medical applications: a survey. IEEE Trans Robot 31 (6):1261–1280

6. Rucker DC, Webster RJ III, Statics and dynamics of continuum robots with general tendon routing and external loading. IEEE Trans Robot 27(6):1033–1044 (2011)

7. Camarillo DB, Milne CF, Carlson CR, Zinn MR, Salisbury JK (2008) Mechanics modeling of tendon-driven continuum manipulators. IEEE Trans Robot 24(6):1262–1273

8. Nguyen T-D, Burgner-Kahrs J (2015) A tendon-driven continuum robot with extensible sections. In: IEEE/RSJ international conference on intelligent robots and systems (IROS), pp 2130–2135

9. Murphy RJ, Kutzer MD, Segreti SM, Lucas BC, Armand M (2014) Design and kinematic characterization of a surgical manipulator with a focus on treating osteolysis. Robotica 32 (6):835–850

10. Cianchetti M, Arienti A, Follador M, Mazzolai B, Dario P, Laschi C (2011) Design concept and validation of a robotic arm inspired by the octopus. Mater Sci Eng C 31 (6):1230–1239

11. Simaan N, Xu K, Wei W, Kapoor A, Kazanzides P, Taylor R, Flint P (2009) Design and integration of a telerobotic system for minimally invasive surgery of the throat. Int J Rob Res 28(9):1134–1153

12. Ding J, Goldman RE, Xu K, Allen PK, Fowler DL, Simaan N (2013) Design and coordination kinematics of an insertable robotic effectors platform for single-port access surgery. IEEE/ASME Trans Mechatron 18 (5):1612–1624

13. Ikuta K, Matsuda Y, Yajima D, Ota Y (2012) Pressure pulse drive: a control method for the precise bending of hydraulic active catheters. IEEE/ASME Trans Mechatron 17 (5):876–883

14. Bailly Y, Amirat Y, Fried G (2011) Modeling and control of a continuum style microrobot for endovascular surgery. IEEE Trans Robot 27(5):1024–1030

15. Chen G, Pham MT, Redarce T (2009) Sensor-based guidance control of a continuum robot for a semi-autonomous colonoscopy. Rob Auton Syst 57(6–7):712–722

16. Rucker DC, Jones BA, Webster RJ III (2010) A geometrically exact model for externally loaded concentric tube continuum robots. IEEE Trans Robot 26(5):769–780

17. Dupont PE, Lock J, Itkowitz B, Butler E (2010) Design and control of concentric-tube robots. IEEE Trans Robot 26(2):209–225

18. Hendrick RJ, Mitchell CR, Herrell SD, Webster RJ III (2015) Hand-held transendoscopic robotic manipulators: a transurethral laser prostate surgery case study. Int J Rob Res 34 (15):1559–1572

19. Swaney PJ, Mahoney AW, Hartley BI, Remirez AA, Lamers EP, Feins RH, Alterovitz R, Webster RJ III (2017) Toward transoral peripheral lung access: combining continuum robots and steerable needles. J Med Robot Res 2 (1):1750001

20. Bergeles C, Gosline AH, Vasilyev NV, Codd PJ, Pedro J, Dupont PE (2015) Concentric tube robot design and optimization based on task and anatomical constraints. IEEE Trans Robot 31(1):67–84

21. Swaney PJ, Gilbert HB, Webster RJ III, Russell PT III, Weaver KD (2015) Endonasal skull base tumor removal using concentric tube continuum robots: a phantom study. J Neurol Surg B Skull Base 76(2):145–149

22. Comber DB, Barth EJ, Webster RJ (2014) Design and control of an magnetic resonance compatible precision pneumatic active cannula robot. J Med Devices 8(1):011003

23. Burgner J, Swaney PJ, Lathrop RA, Weaver KD, Webster RJ III (2013) Debulking from within: a robotic steerable cannula for intracerebral hemorrhage evacuation. IEEE Trans Biomed Eng 60(9):2567–2575

24. Butler EJ, Hammond-Oakley R, Chawarski S, Gosline AH, Codd P, Anor T, Madsen JR, Dupont PE, Lock J (2012) Robotic neuroendoscope with concentric tube augmentation. In: IEEE/RSJ international conference on intelligent robots and systems (IROS), pp 2941–2946

25. Bodani V, Azimian H, Looi T, Drake J (2014) Design and evaluation of a concentric tube robot for minimally-invasive endoscopic paediatric neurosurgery. Hamlyn Symp Med Robot 1(1):25–26

26. Hendrick RJ (2017) System design and elastic stability modeling of transendoscopic continuum robots. Ph.D. dissertation

27. Swaney PJ, York PA, Gilbert HB, Burgner-Kahrs J, Webster RJ III (2016) Design, fabrication, and testing of a needle-sized wrist for

surgical instruments. ASME J Med Devices 11 (1):014501

28. Burgner J, Rucker DC, Gilbert HB, Swaney PJ, Russell PT III, Weaver KD, Webster RJ III (2014) A telerobotic system for transnasal surgery IEEE/ASME Trans Mechatron 19 (3):996–1006

29. Camarillo DB, Krummel TM, Salisbury JK Jr (2004) Robotic technology in surgery: past, present, and future. Am J Surg 188(4):2–15

30. Wirz R, Torres L, Swaney PJ, Gilbert HB, Alterovitz R, Webster RJ III, Weaver KD, Russell PT III (2015) An experimental feasibility study on robotic endonasal telesurgery. Neurosurgery 76(4):479–484

31. Godage IS, Remirez AA, Wirz R, Weaver KD, Burgner-Kahrs J, Webster RJ III (2015) Robotic intracerebral hemorrhage evacuation: an in-scanner approach with concentric tube robots. In: IEEE/RSJ international conference on intelligent robots and systems (IROS), pp 1447–1452

32. Gilbert HB, Neimat J, Webster RJ III (2015) Concentric tube robots as steerable needles: achieving follow-the-leader deployment. IEEE Trans Robot 31(2):246–258

33. Wieser HG, Ortega M, Friedman A, Yonekawa Y (2003) Long-term seizure outcomes following amygdalohippocampectomy. J Neurosurg 98(4):751–763

34. Comber DB, Pitt EB, Gilbert HB, Powelson MW, Matijevich E, Neimat JS, Webster RJ III, Barth EJ (2017) Optimization of curvilinear needle trajectories for transforaminal hippocampotomy. Oper Neurosurg 13(1):15–22

35. Pitt EB, Comber DB, Chen Y, Neimat JS, Webster RJ, Barth EJ (2016) Follow-the-leader deployment of steerable needles using a magnetic resonance-compatible robot with stepper actuators. J Med Devices 10 (2):020945

36. Xu K, Simaan N (2008) An investigation of the intrinsic force sensing capabilities of continuum robots. IEEE Trans Robot 24 (3):576–587

37. Okamura AM (2009) Haptic feedback in robot-assisted minimally invasive surgery. Curr Opin Urol 19(1):102

38. Mahoney A, Bruns T, Swaney PJ, Webster RJ III (2016) On the inseparable nature of sensor selection, sensor placement, and state estimation in continuum robots or 'Where to put your sensors and how to use them'. In: IEEE international conference on robotics and automation (ICRA), pp 4472–4478

39. Kim B, Ha J, Park FC, Dupont PE (2014) Optimizing curvature sensor placement for fast, accurate shape sensing of continuum robots. In: IEEE international conference on robotics and automation (ICRA). IEEE, Piscataway, pp 5374–5379

40. Csencsits M, Jones BA, McMahan W, Iyengar V, Walker ID (2005) User interfaces for continuum robot arms. In: IEEE/RSJ international conference on intelligent robots and systems (IROS), pp 3123–3130

41. Travaglini TA, Swaney PJ, Weaver KD, Webster RJ III (2015) Initial experiments with the leap motion as a user interface in robotic endonasal surgery. In: IFTOMM international symposium on robotics & mechatronics, pp 171–179

42. Anor T, Madsen JR, Dupont P (2011) Algorithms for design of continuum robots using the concentric tubes approach: a neurosurgical example. In: 2011 IEEE international conference on robotics and automation (ICRA). IEEE, Piscataway, pp 667–673

43. Burgner J, Gilbert HB, Webster RJ III (2013) On the computational design of concentric tube robots: incorporating volume-based objectives. In: IEEE international conference on robotics and automation, pp 1185–1190

44. Torres LG, Webster RJ, Alterovitz R (2012) Task-oriented design of concentric tube robots using mechanics-based models. In: IEEE/RSJ international conference on intelligent robots and systems (IROS), pp 4449–4455

45. Ha J, Park FC, Dupont PE (2014) Achieving elastic stability of concentric tube robots through optimization of tube precurvature. In: IEEE/RSJ international conference on intelligent robots and systems (IROS), pp 864–870

46. Baykal C, Torres LG, Alterovitz R (2015) Optimizing design parameters for sets of concentric tube robots using sampling-based motion planning. In: IEEE/RSJ international conference on intelligent robots and systems (IROS), pp 4381–4387

47. Hendrick RJ, Gilbert HB, Webster RJ (2015) Designing snap-free concentric tube robots: a local bifurcation approach. In: IEEE international conference on robotics and automation (ICRA). IEEE, Piscataway, pp 2256–2263

Chapter 4

Shared-Control Robots

Christopher J. Payne, Khushi Vyas, Daniel Bautista-Salinas, Dandan Zhang, Hani J. Marcus, and Guang-Zhong Yang

Abstract

This chapter reviews shared-control robots, a class of robotic device in which the surgeon and the robot simultaneously manipulate the surgical tool together. The shared-control approach seeks to exploit the superior aspects of humans and machines, to enable more precise interventions while ensuring the human surgeon retains executive control. Much of the technology discussed in this chapter is emerging research and many of the described systems have been developed for generic microsurgical interventions. Nonetheless, the broad concepts behind these surgical systems are highly applicable to neurosurgery and particularly to microsurgical procedures. We start by presenting an exemplar of a grounded, shared-control robot: the *Steady-Hand* system. We then review a series of handheld smart surgical devices, including *Micron*, a handheld tremor cancellation device. This chapter also presents handheld devices capable of augmenting haptic feedback to surgeons performing delicate neurosurgical tasks, image-guided handheld devices with embedded robotic actuation, and a new generation of handheld microscopic imaging devices for visualizing tumors.

Key words Medical robotics, Surgical robotics, Shared-control robots, Handheld robots, Medical devices, Smart surgical devices

1 Introduction

Neurosurgical interventions carry an intrinsic risk of morbidity and mortality. The physical structure of the human brain is extremely complex and often challenging to navigate. Furthermore, neurovascular tissue is exquisitely prone to injury and has limited capacity for regeneration. To work in a safe and effective manner, neurosurgeons must operate with precision and delicateness. Advances in neurosurgical technology has certainly sought to improve the former: the adoption of operating microscopes and fine surgical instrumentation has greatly advanced surgical precision. More recently, computer-based image-guidance systems have significantly advanced the ability of surgeons to localize lesions hidden within the brain. Nonetheless, there are inherent limitations to the precision with which a human surgeon can operate; and herein lies

Hani J. Marcus and Christopher J. Payne (eds.), *Neurosurgical Robotics*, Neuromethods, vol. 162,
https://doi.org/10.1007/978-1-0716-0993-4_4, © Springer Science+Business Media, LLC, part of Springer Nature 2021

the opportunity for robotics. Unlike humans, robotic manipulators can move with unparalleled precision. It is a pursuit for precision that has driven many of the developments in the wider field of medical robotics. For the field of neurosurgery, the ability to operate with enhanced precision may prove particularly impactful in enhancing patient outcomes. Although technological trends have tended to focus on improving spatial precision in neurosurgery, robotic devices may also allow surgeons to perform interventions on the brain with greater delicateness as well. Recent studies have quantified the surgical interaction forces in neurosurgical procedures [1–3] which are often at magnitudes below that of human touch perception. While such surgical manipulation forces are below what a human can feel, robotic systems can be designed to sense miniscule surgical forces and feed back these delicate haptic cues to enhance the force perception of an operating neurosurgeron. Even more critically, studies have quantified the force thresholds at which potential injury to neurovascular tissue is likely to occur [1]. While a human surgeon cannot objectively distinguish such a force threshold during a procedure, surgical robotic systems can impose force limits to prevent iatrogenic injury.

One of the very first medical robotic systems reported in the literature was a stereotactic positioning system for neurosurgery [4]. Like many of the other pioneering efforts in medical robotics, this system was an industrial robot repurposed for a medical application. These early medical robots, like their industrial counterparts, were designed to perform tasks autonomously. Their operation was based on preoperative programming of the robot, and with minimal intraoperative input from the surgeon. However, autonomous robots executing predefined trajectories and commands are not always well suited for conducting complex neurosurgical procedures. Instead, teleoperated robots, which have no autonomy [5], are generally preferred. In such teleoperated robotic systems, the surgeon operates a human-machine interface at a control console and these input motions are then replicated on a robotic device that operates on the patient. In a teleoperated system, the surgeon retains complete control of the robotic device during the procedure. This approach has several advantages; it is possible to filter out tremulous input motion, scale motion between the console interface and the remote robotic system; and improve the surgeon's ergonomics. A major drawback of this approach is that it warrants a significant amount of hardware: a console interface, a robotic manipulator, and a complex control system. These systems are expensive, physically large and increase the procedure setup time. To address these challenges, innovators are looking to alternative approaches that empower surgeons with robotic assistance.

In a shared-control system, both the robot and the surgeon directly manipulate the surgical tool. Like a teleoperated system, this assistive approach fuses together the best of humans and robots. However, since the surgeon now manipulates the robot directly rather than through a remote console, there is inherently less hardware warranted in such a setup. Thus, shared-control robots can have a significantly smaller footprint and can be innately less expensive to manufacture. Shared-control robots allow the surgeon to directly hold the surgical instrumentation at the patient-side, in a manner they are already accustomed to from their clinical training. Shared-control robots can be further stratified into grounded and ungrounded systems. Grounded robots are mechanically affixed to a grounding frame such as a mobile cart, or some other form of anchor. Whereas ungrounded robots are handheld and incorporate robotic features within the surgical tool itself. This latter class of device represents an even greater degree of miniaturization and potential cost reduction. Furthermore, these handheld "smart" devices can be seamlessly integrated in and out of the surgical workflow as and when they are required, with little or no setup time. Like the wider mobile computing revolution which has seen considerable miniaturization of computing technology, smart handheld surgical tools have significant potential to increase the adoption of robotic technology around the world [6]. This trend toward smaller, smarter surgical instrumentation is highlighted in Fig. 1. A summary of shared-control robotic system features, pros, cons, and examples of use are reported in Table 1.

2 Grounded Shared-Control Robots

Grounded shared-control robots are fundamentally based on the concept that both the surgeon and robot physically manipulate the surgical tool together. Fusing together the best aspects of humans and robots, these systems can empower surgeons with superhuman operating precision and haptic perception. This paradigm has found significant success in the orthopedics field where robotic manipulators assist surgeons to precisely sculpt bones for implant preparation [7]. This approach can also have significant impact within the microsurgical domain. In this section, we illustrate the concepts of a grounded shared-control system through an exemplar device known as the Steady-Hand system [8].

The Steady-Hand system can improve both the positional precision with which a surgeon can manipulate a surgical tool, as well as the surgeon's haptic perception of the tool–tissue interactions during surgery (*see* Fig. 2). The robotic manipulator holds a surgical tool and incorporates multi degree-of-freedom (DoF) force sensing at the tool handle which the surgeon can manipulate. Using a control scheme based on this force input, the robotic system can sense the surgeon's intent and adjust the position of the surgical tool to follow that which is desired by the surgeon. The

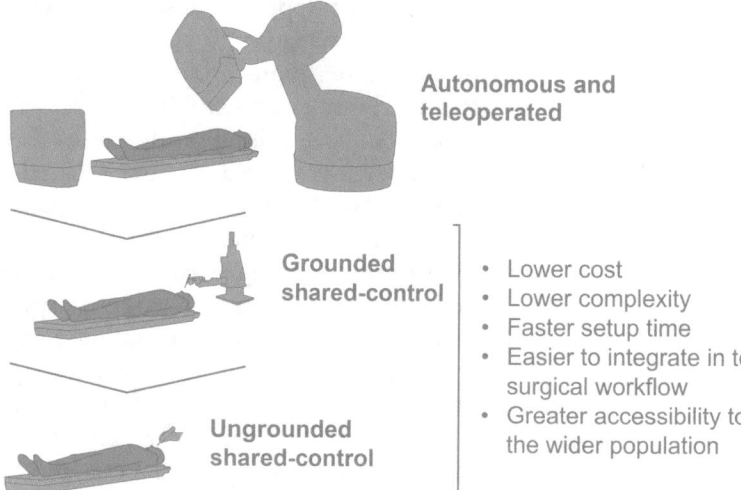

Autonomous and teleoperated

Grounded shared-control

- Lower cost
- Lower complexity
- Faster setup time
- Easier to integrate in to surgical workflow
- Greater accessibility to the wider population

Ungrounded shared-control

Fig. 1 Trends in medical robotics toward smaller, lower cost systems echo that of the mobile computing revolution [6]

robotic control system can be programmed to provide a tunable level of mechanical resistance to help steady the surgical tool. By tuning the robot dynamics, this resistive effect can help damp out tremulous motion and allow smooth positioning of the surgical instrument. It is also possible for the Steady-Hand robot to provide resistive forces based on the surgical interaction forces at the tool tip. Like a telerobotic system, the Steady-Hand can be programmed to provide haptic feedback based on the forces experienced by the surgical tool. Given the low magnitude forces warranted in micro-surgery, it is typical that these surgical interaction forces are magnified through the robotic system in order to augment the surgeon's haptic perception during an intervention.

In addition to steadying the surgical tool position and controlling the force applied, another means by which the Steady-Hand can potentially improve patient safety is by constraining the surgical tool away from anatomically-critical regions. The "virtual fixture" or "active constraint" concept in medical robotics provides a means of constraining the robot to a virtually defined point, trajectory or spatial region. In the same way that a ruler can constrain a pen to a straight line, a shared-control robot can incorporate preprogrammed constraints to allow safe navigation of the surgical tool during an intervention. The robot can allow free motion of the surgical tool until the constraint is approached; at this point, the robot resists the surgeon to prevent the surgical tool moving beyond the preprogrammed constraint [9]. This concept requires coregistration of the robot coordinate system with the real-world anatomy. Such a system may facilitate safer positioning of surgical instruments, particularly in keyhole and microsurgical procedures where it is not always possible to visualize the entirety of the surgical tool at a given instance.

Table 1
Summary for the features, pros, and cons of different approaches for shared-control robots in neurosurgery

Class of shared-control robotic system	Key features	Pros	Cons	Examples of use
Grounded-shared-control	• Surgeon and robotic assembly physically manipulate a surgical tool together • Robotic assembly is anchored to a grounding frame	• Ability to suppress tremulous hand motions • Can incorporate force scaling functionality • Can incorporate active constraints for tool guidance • Surgical tool can be locked into position to act as a "third arm"	• Relatively large physical footprint required • Some workspace constraints due to mechanical assembly • Relatively complex hardware required • Requires setup procedure	• Microsurgical interventions warranting fine motions and very delicate force applications • Interventions in close proximity to anatomically critical areas
Ungrounded-shared-control robots	• Robotic features are integrated to a handheld device that has no external grounding frame • Devices exploit the existing dexterity of surgical operator	• Can be seamlessly integrated in and out of the surgical workflow • Physically compact • Unobtrusive to manipulate • Can be made very simple and potentially at a relatively low cost	• Individual tools have specific functionality and are less versatile • Optically tracked handheld devices have workspace constraints and require setup	• Tremor-suppression devices can be used in microsurgical applications that require fine tool manipulation • Force-feedback devices can aid force perception or warn of dangerous force levels in delicate microsurgical tasks • Image-guided procedures requiring enhanced precision • Emerging applications involve in situ microscopic imaging of tissues for precise tumor resection

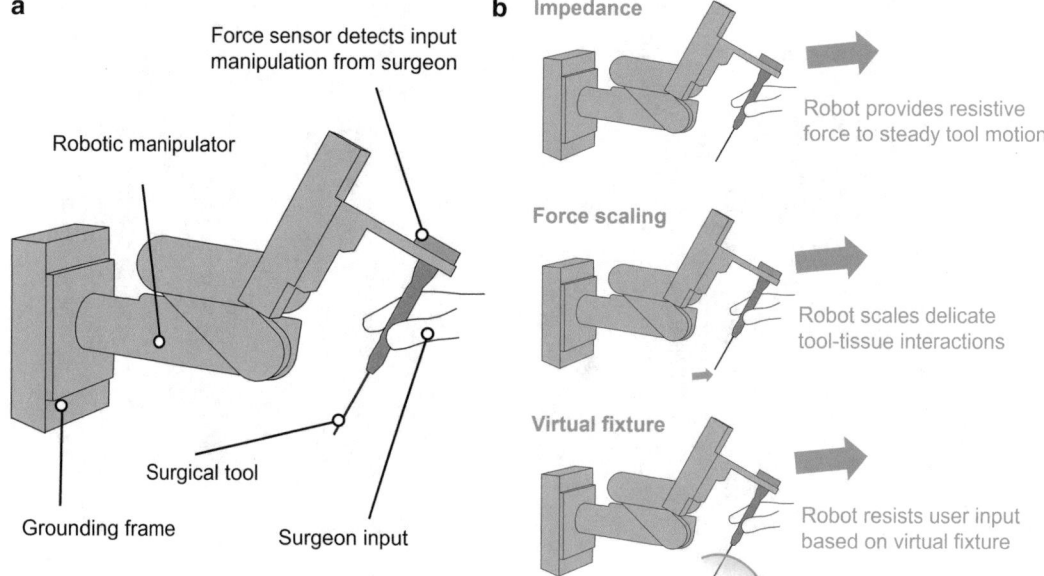

Fig. 2 (**a**) Schematic showing the Steady-Hand concept as an exemplar of a grounded shared-control robot [8]. (**b**) Control paradigms for the Steady-Hand system [8, 9]

3   Ungrounded Shared-Control Robots

We have reviewed how grounded shared-control robots can steady tremulous motion, augment haptic perception and constrain a surgical tool to move in a predefined region or pathway. Nonetheless, this type of robot still has some drawbacks that have limited adoption of these concepts into clinical practice. Firstly, grounded robotic manipulators have an inherently large physical footprint as they are attached to a grounding frame and like manufacturing robots, they possess a degree of rigidity and physical mass to deliver the precise positioning required for microsurgery. Furthermore, such robots need to be set up in the operating room which can not only take time but also disrupt the surgical workflow. An additional disadvantage of using such a robot is the limited physical workspace of the robotic manipulator arm.

Ungrounded shared-control robots are held by the hands of the surgeon and remain unattached to an external grounding frame. This class of shared-control robotic device is broad and such devices can vary in their function and application, but the common idea is to integrate robotic features into existing surgical instrumentation to make smart devices with a small physical footprint [10]. For many surgical applications, the human arm and hand is a very effective gross manipulator, and the integrated robotic functionality within the handheld device can serve to augment surgical precision at the tool tip or provide haptic feedback to the surgeon. Handheld smart devices can be more cost-effective compared to grounded robotic systems, often warranting less hardware to achieve the same goal. Such devices already possess a degree of familiarity to the operating surgeon; and hand-held tools can be seamlessly integrated into the surgical workflow when required. Four categories of smart handheld device are discussed in this section: (1) tremor suppression, (2) haptic feedback devices, (3) active image-guidance systems, and (4) smart microscopy imaging.

3.1 Tremor Suppression Handheld Devices

One of the key benefits of shared-control robots, such as the Steady-Hand system is the ability to filter out inadvertent motions such as physiological tremor. The medical usage of the term tremor is defined as a rapid quasi-periodic motion [11]. In the context of tremor suppression handheld devices, the term *tremor* can be more broadly applied to refer to involuntary hand motions that cause positioning errors. These involuntary motions are a superposition of slow drifting motions, sporadic fast jumping motions and higher frequency physiological tremor. The total magnitude of such involuntary hand motions can be up to a few hundred microns [12]. In microsurgical interventions, some surgical maneuvers can be of a similar physical scale to that of these involuntary tremulous hand motions. Ungrounded, handheld robotic systems represent an alternative architecture for cancelling involuntary hand motion in surgery. In a grounded system, a relatively large robot, with multiple linkages is anchored to the ground and the surgeon manipulates the end-effector where the surgical tool is located. In an ungrounded tremor cancelling device, the surgeon holds the base handle of the surgical tool and a miniaturized micromanipulation platform embedded within the device itself can servo the surgical tool tip independently of the operating surgeon. Tremor suppression can be achieved using a threefold approach. Firstly, the device senses the surgeon's input motion. Secondly, a controller filters out the intended motion from the erroneous motion. Thirdly, the on-board actuation system servos the surgical tool tip in the opposite direction to that of the tremor to effectively cancel out the erroneous motion. A number of handheld tremor suppression devices have been developed [13–16], including *Micron*, a device developed at Carnegie Mellon University [12, 17, 18]. Most of the research into these devices has focused on (1) technologies to effectively sense the physiological tremor input, (2) motion filtering methodologies, and (3) actuation strategies.

Early tremor suppression devices utilized inertial sensors, such as those used in smartphones, to sense the motion of the device. An advantage of this approach is that the sensing system is fully integrated within the device, but the sensing fidelity of such an approach is limited. Inertial sensors can effectively measure accelerations but are less accurate in localizing the device's spatial

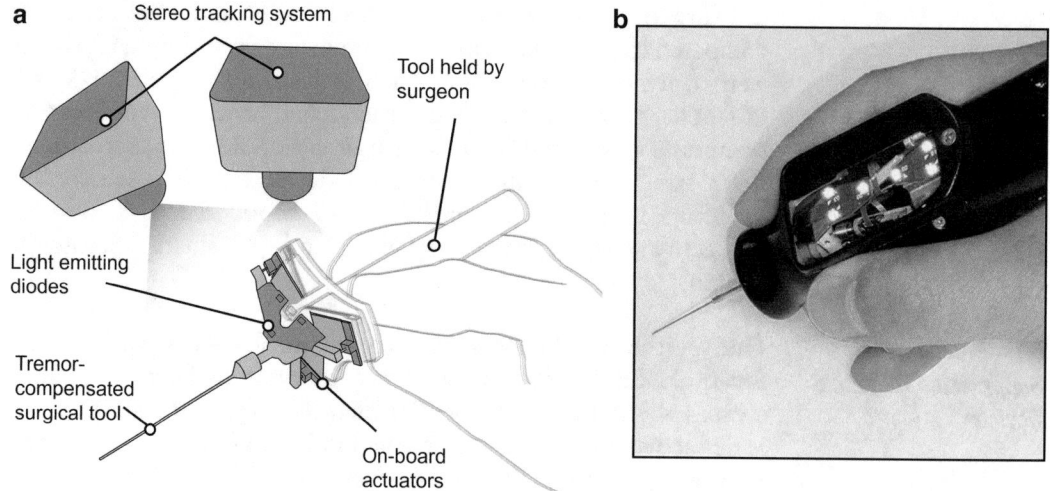

Fig. 3 (**a**) Schematic showing the principle of a three-DoF tremor suppression handheld device utilizing optical tracking [12, 19]. (**b**) Photograph of the six-DoF Micron device (with permission from SAGE Publications Inc. [20])

position. A more accurate approach is to use an image-guidance system based on optical tracking; such systems can determine the tool position to within a ~4μm resolution and at the high frame-rates required for effective tremor suppression [19]. Such optical tracking systems consist of two cameras capable of tracking a set of light emitting diode (LED) markers that are mounted on the device (Fig. 3a). Furthermore, it is possible to use a camera to provide an additional sensing input to allow for higher-level control functions based on the surgical scene. Examples of this approach include a "snap-to function" that can automatically guide the surgical tool to a predetermined area of interest, as well as implementation of virtual fixtures to guide the tool away from anatomical regions, or to guide the tool along a prespecified path [21–24].

Unlike grounded robotic systems which typically adopt conventional actuators (such as electric motors and gear drives), the actuation systems for handheld devices have fundamentally contrasting requirements. Actuators need to be compact, lightweight and capable of rapid operation in order to effectively cancel high-frequency, tremulous motion. For this reason, tremor suppression systems such as Micron typically use piezoelectric actuators that work by physically deforming in order to steer the surgical tool. A drawback to this approach is that the physical travel range of such actuators is limited, and so only very fine tremulous motions can be compensated. Additionally, many tremor suppression devices have three DoFs to compensate for 3D position inaccuracies but are not effective in correcting tool orientation errors. To overcome this limitation, the most recent embodiment of Micron utilizes miniature linear actuators with six DoFs (Fig. 3b). User studies with these tremor suppression systems have demonstrated their efficacy in a multitude of scenarios from path tracking tasks to clinical simulations [18, 25]. In one circle tracing task example, tremor was effectively suppressed with a 90% reduction in positioning error when use of the active device is compared to a freehand control [18].

3.2 Handheld Devices with Enhanced Haptic Feedback

Neurosurgical robotic systems have commonly focused on augmenting positional accuracy where the goal is to manipulate the surgical tool with greater precision. Less attention has been paid to methodologies for improving the delicateness with which neurosurgical interventions are performed. Robotic systems can move more precisely than humans, but they can also be designed to sense tool–tissue forces with a fidelity far superior to that of a human surgeon. In neurosurgery, iatrogenic injury is often irreversible and can have severe consequences to the patient, so there has been recent interest in smart surgical systems that can reduce the forces being exerted on the brain during a procedure. The Steady-Hand robot incorporates force-feedback features for augmenting haptic feedback by scaling the minute surgical interaction forces to higher magnifications that can be better perceived by the surgeon operator. In grounded, shared-control robots such as the Steady-Hand, force scaling is made possible because the robotic manipulator can resist the surgeon and the resulting reaction forces are supported by the grounding frame of the robot. While it would be advantageous to incorporate this ability into lower cost, smaller footprint handheld devices, the lack of an external grounding frame makes this challenging. Researchers have considered some handheld device methodologies to overcome this challenge [26–28]. Stetten et al. devised a device which provides bracing to the back of the surgeon's hand in order to support a reaction force and enable force magnification functionality [26] (*see* Fig. 4a). A force sensor integrated within the device measures the axial forces at the surgical tool tip and a linear actuator then exerts a scaled force on to the shaft of the handheld device. User studies have demonstrated that this approach can improve the force-sensing threshold of the human operator. This ability to feel surgical forces below the normal human threshold could allow critical neurosurgical interventions to be performed more delicately. Researchers have also considered designs which provide force-feedback directly to the surgeon's fingertip [28] (*see* Fig. 4b) as well as forceps designs that can relay back the tissue grasping force at scaling factors of up to ×50 [32].

Handheld devices with force-scaling functions may improve the delicateness with which surgery is carried out, with the goal of reducing iatrogenic injury. An alternative formulation of this problem is to understand the force thresholds at which iatrogenic injury to brain tissue occurs, and to limit tool–tissue interactions beyond

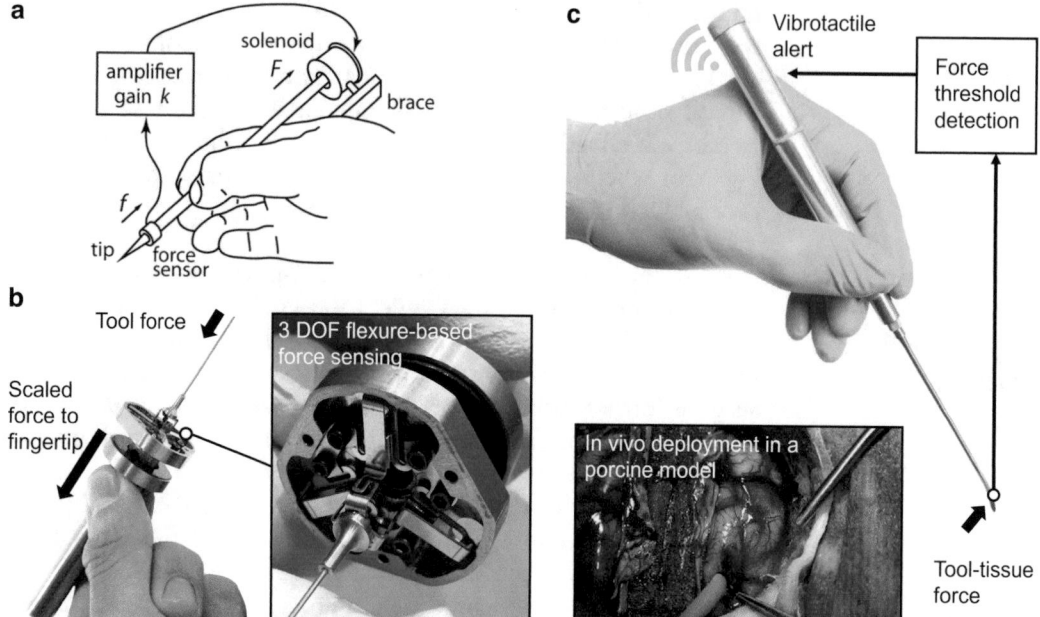

Fig. 4 (**a**) Diagram illustrating the force magnifying concept in a handheld tool (©2011 reproduced from Stetten et al. with permission from Springer). (**b**) Photographs showing the embodiment of a handheld force-amplifying device [29] (with permission from Christopher J. Payne and the Hamlyn Centre for Robotic Surgery, Imperial College London). (**c**) Haptic feedback microdissector (top) [30] (©2015 reproduced from Payne et al. with permission from Springer); and the haptic microdissector device deployed in vivo (bottom) (reproduced in accordance with the Creative Commons Attribution License [31])

these levels. Researchers have investigated the magnitude of forces experienced during neurosurgery [2, 3] as well as the force thresholds at which injury occurs in dissection procedures [1]. Forces in neurosurgical dissection are typically below 1 N, and injury can occur at thresholds as low as 280 mN. We have developed a smart microdissection device that senses delicate surgical interaction forces and provides vibrotactile feedback at a preprogrammed force threshold [30]. In a series of psychophysical user studies, this device was demonstrated to limit force application when users were distracted, or if their nominal haptic perception was distorted. Subsequently, it was shown that the device can significantly improve the dissecting performance of trainee surgeons in validated dissection tasks, bringing their operating delicateness in line with that of expert and intermediate-level surgeons, without impacting other performance metrics [31]. More recently, researchers have also developed a soft, silicone-based retracting device that uses a fluidic system to infer retraction force and provide visual force feedback to the surgeon [33]. In addition to being compact, low-cost devices, these smart tools serve to guide the surgeon so that they remain free to use their own judgment when executing a procedure. Such devices may be particularly useful in training

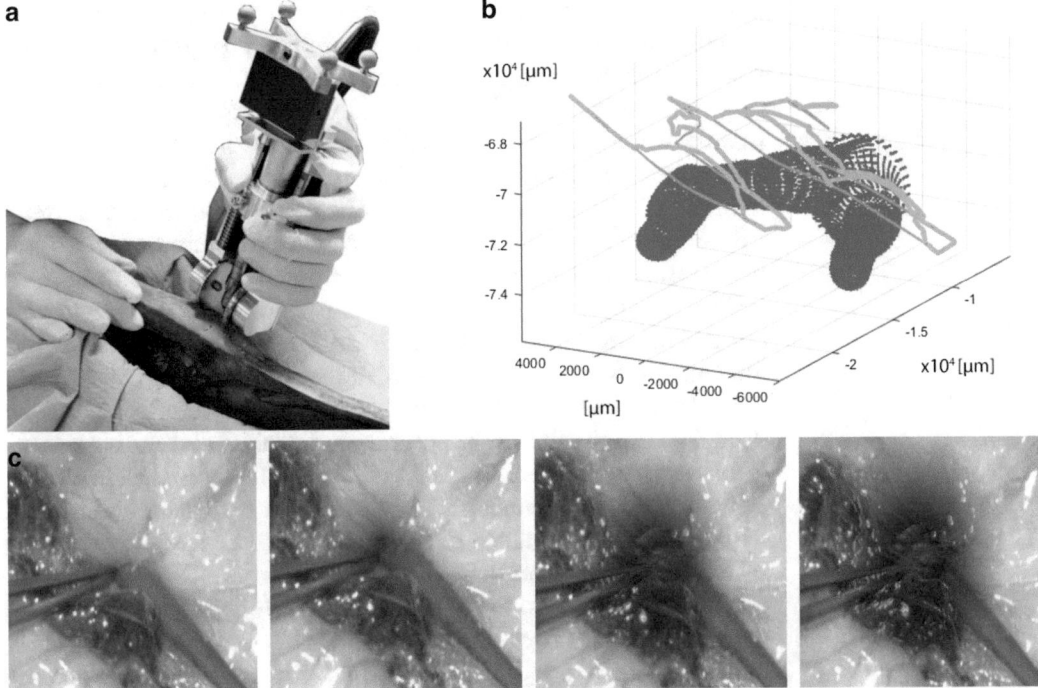

Fig. 5 (**a**) Craniostar device for performing image-guided craniotomy (©2009 reproduced from Kane et al. with permission from Springer [34]). (**b**) Example of active-constraints applied to neurosurgical anatomy using Micron (blue structure represents a reconstructed blood vessel, and the orange trace is the surgical tool path without active constraints; the green trace is the surgical tool path with active constraints applied) (©2018 reproduced from Moccia et al. with permission from Springer [35]). (**c**) Visual force feedback using a smart microdissector as applied to a brain phantom with visualization of sub-surface anatomy (©2015 reproduced from Gras et al. with permission from Springer [36])

junior surgeons to learn the safe dissection force-levels during neurosurgical procedures.

3.3 Smart Devices for Image-Guided Neurosurgery

Image-guidance is ubiquitous in neurosurgery around the world. The ability to guide surgical tools based on preoperative medical imaging has greatly enhanced surgical precision. The utility of such image-guided approaches can be further enhanced by integrating active robotic features into handheld devices. For example, the Craniostar [34] is an image-guided handheld mobile robot that allows for semiautomated craniotomy. The handheld device is positioned on the skull and is spatially localized using optical tracking. The device incorporates actuated wheels which help guide the craniotomy trajectory as the surgeon manipulates the device over the skull (Fig. 5a). In another example, a handheld tool was developed with a retractable tip to allow for the implementation of virtual fixtures [37]. As the device approaches an anatomically-critical region, the surgical tool tip can retract to prevent an inadvertent collision. This device only requires a single actuator integrated within the shaft of the tool to provide virtual fixture functionality at a lower complexity compared to grounded robotic systems that require multiple linkages. The device also incorporates ungrounded haptic feedback so that the user can feel the virtual fixture when the device makes contact. Virtual fixtures with ungrounded handheld devices have been directly investigated for neurosurgical applications [35]. In this work, the aforementioned Micron device was used to assist petroclival meningioma resection by suppressing tremor and using active image guidance to prevent inadvertent tool collisions with the surrounding major blood vessels in a phantom model. Results from this study, shown in Fig. 5b, illustrate how Micron can servo the surgical tool tip away from critical blood vessels. Active image-guidance systems can also be used in combination with augmented reality methodologies to inform surgeons of subsurface anatomical features during dissection tasks. Using a force-sensing neurosurgical dissector combined with an image-guidance system, visual force-feedback can be relayed to the surgeon in real-time. In this work, force information was incorporated into the surgical scene by highlighting the region where the dissector contacts the tissue while simultaneously providing an overlay visualization of the anatomical structures beneath the surface [36]. The size and intensity of the augmented reality overlay can be modulated according to the force experienced by the dissector (Fig. 5c).

3.4 Smart Handheld Devices for In Vivo Microscopic Imaging

The chapter so far has reviewed various advances in smart handheld robotic devices, with many devices designed to improve the precision of neurosurgical interventions. Another emerging opportunity to enhance surgical precision is through intraoperative imaging tools capable of examining the microarchitecture of the brain to improve surgical resection margins. One of the challenges faced by neurosurgeons in the operating room is the inability to recognize indistinguishable pathologically altered tissue from healthy brain parenchyma, which can often result in positive margins and tumor recurrence. Confocal laser endomicroscopy (CLE) is an intraoperative imaging tool which can provide live, high-resolution imaging of tissue morphology, cytoarchitecture, and intracellular elements at various depths within the tissue [38]. When combined with intravenous and topical fluorophores, CLE systems can deliver real-time information about brain neoplasms at the cellular level during an intervention.

Although still in an early stage of clinical adoption, bespoke research-based and commercial CLE systems have demonstrated the capability to rapidly differentiate different brain tissue types during resection, in animal and human studies [39–42]. The commercial Optiscan CLE system was first used to image glioblastomas in animal models with a lateral resolution close to 0.7 μm. Using fluorescein and acriflavine hydrochloride staining agents delivered

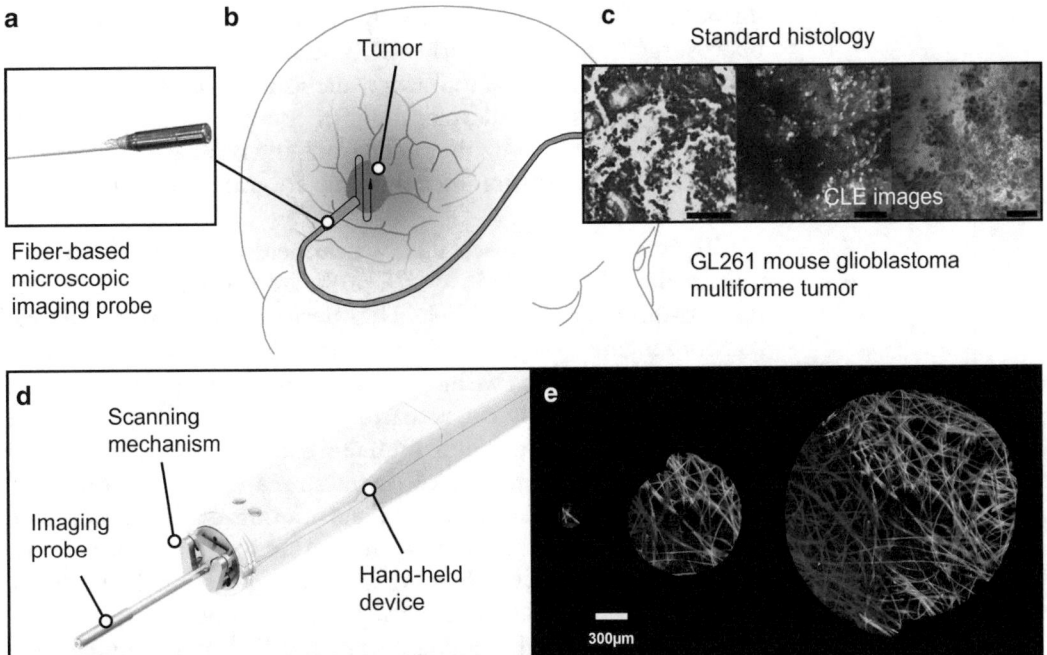

Fig. 6 (**a**) Cellvizio CLE fiber-based imaging probe (with permission from Khushi Vyas and the Hamlyn Centre for Robotic Surgery, Imperial College London). (**b**) Intraoperative surgical imaging with a CLE-scanning device can be used to delineate tumor margins. (**c**) Standard histologic and confocal optical sections taken from the same region of a typical GL261 mouse glioblastoma multiforme tumor, scale bar is 250 μm (image used with permission from Sankar, Tejas, Delaney, Peter M., Miniaturized Handheld Confocal Microscopy for Neurosurgery: Results in an Experimental Glioblastoma Model, Neurosurgery, 2010, volume 66, issue 2, Pages 410–418, and by permission of Oxford University Press [39]). (**d**) Handheld robotic CLE scanning device (image used with permission from Christopher J. Payne and the Hamlyn Centre for Robotic Surgery, Imperial College London [44]). (**e**) Sequential spiral mosaicking using a handheld robotic device (images show composition of mosaic over time), scale bar is 300 μm (image used with permission from Christopher J. Payne and the Hamlyn Centre for Robotic Surgery, Imperial College London)

intravenously and topically, researchers were able to distinguish between tumor and normal regions using a handheld confocal microscope [39]. Recently, a commercial Cellvizio® 100 series CLE with CranioFlex™ confocal Miniprobes™ from Mauna Kea Technologies has obtained FDA clearance for neurosurgical applications. Such systems can provide lateral resolutions down to 1 μm, for a field-of-view of 240 μm through a small flexible probe (2.6 mm diameter).

Despite the promise of such technology, existing CLE systems have limitations that have prevented clinical adoption. Firstly, the small size and probe flexibility causes difficulties in maintaining adequate probe-tissue contact and manipulation dexterity, especially during handheld manual scanning applications. Furthermore, a major limitation of existing CLE systems is that the field-of-view is limited by the size of the fiber-bundle, typically 0.24–0.8 mm diameter. This makes it extremely challenging for the operating neurosurgeon to interpret the CLE images correctly and acquire information about the lesion sites while scanning large areas such as resection cavities. The development of robotically actuated hand-held devices for manipulating micro imaging probes for automatic, consistent probe positioning and tissue coverage can address these shortcomings.

To improve probe stability, a handheld assistive robotic tool was developed to maintain a CLE probe in consistent contact with the surrounding tissue [43]. This device works using a force-control scheme to regulate probe contact force during imaging. A force sensor integrated within the device monitors the probe–tissue contact force and a linear actuator can servo the imaging probe back-and-forth to track and regulate contact force. This system could compensate physiological disturbances, either from the tissue being imaged, or from the handheld motion of the surgeon. In ex vivo experimentation, it was shown that tissue imaging consistency was significantly enhanced when using the handheld device versus manual manipulation of the imaging probe.

Given the limited imaging size of the CLE fiber-bundles, the ability to rapidly and precisely image large regions of tissue will be critical for the translation of CLE technology into the clinic. A robotic handheld device recently developed by the Hamlyn Centre for Robotic Surgery was designed to reliably servo imaging probes with microscale precision to meet this challenge [44]. The clinical-use vision for this device is illustrated in Fig. 6. A CLE probe can be passed through the bore of the handheld device shaft which can be deflected by a robotic micromanipulation system. Micro electric motors are used to actuate lever assemblies which exert lateral deflections to the imaging probe. By controlling the respective motor positions, the optical probe can be programmed to perform a variety of different scanning motions, such as back-and-forth raster motions and spiraling trajectories. As the probe is servoed through a known trajectory, the corresponding images can be stitched together using a technique known as image mosaicking. This methodology allows relatively large regions of tissue to be rapidly scanned and imaged as a whole region. The developed device has a total workspace scan area of 14 mm^2 and when combined with a high frame rate line-scan endomicroscopy system [45], the device can generate high-resolution images over an area of 3 mm^2 in approximately 10 s. The device also incorporates a laser ablation system so that tissue regions can be marked after a scan is complete. With successive scans, the tumor boundary can be mapped out and subsequently resected. More recently, this handheld device has been integrated with a grounded shared-control robot to enable large-area scanning applications where greater workspace is required [46].

4   Conclusions

Shared-control medical robots are a paradigm in which both the surgeon and a robotic system manipulate a surgical tool. Such robotic systems are designed to exploit the respective advantages of humans and robots with the goal of improving surgical precision or delicateness. In neurosurgery, the ability to operate with enhanced precision and delicateness could significantly improve clinical outcomes. This chapter has reviewed two subcategories of shared-control robots: grounded systems that use a robotic manipulator that is affixed to the ground or surgical cart; and ungrounded, smart handheld devices in which the robotic features are integrated into the handpiece of the device itself. We have seen how these devices can aid surgical procedures in a multitude of ways. Surgical precision can be augmented by suppressing tremor and the use of virtual fixtures to constrain the surgical tool away from anatomically-critical regions. The delicateness of procedures can be enhanced by magnifying and limiting tool–tissue forces as the surgeon interacts with the brain. Emerging in vivo microscopy tools can also be combined with handheld surgical robots to provide real-time diagnostic information to neurosurgeons performing tumor resections. A key advantage of shared-control robots, and particularly handheld smart devices, is that they are inherently less complex, physically smaller and often inherently safer. The corollary to this is that such devices can be manufactured at a low cost and integrated seamlessly in and out of the surgical workflow. This in turn may encourage the uptake of robotic systems in neurosurgery, allowing robotics to be more widely accessible to the global population.

References

1. Marcus HJ, Zareinia K, Gan LS et al (2014) Forces exerted during microneurosurgery: a cadaver study. Int J Med Robot Comput Assist Surg 10:251–256

2. Gan LS, Zareinia K, Lama S et al (2015) Quantification of forces during a neurosurgical procedure: a pilot study. World Neurosurg 84:537–548

3. Zareinia K, Maddahi Y, Gan LS et al (2016) A force-sensing bipolar forceps to quantify tool-tissue interaction forces in microsurgery. IEEE/ASME Trans Mechatron 21:2365–2377

4. Kwoh YS, Hou J, Jonckheere EA et al (1988) A robot with improved absolute positioning accuracy for CT guided stereotactic brain surgery. IEEE Trans Biomed Eng 35:153–160

5. Yang GZ, Cambias J, Cleary K et al (2017) Medical robotics-regulatory, ethical, and legal considerations for increasing levels of autonomy. Sci Robot 2:2–4

6. Marcus H, Nandi D, Darzi A et al (2013) Surgical robotics through a keyhole: from today's translational barriers to tomorrow's "disappearing" robots. IEEE Trans Biomed Eng 60:674–681

7. Jakopec M, Harris SJ, Rodriguez y Baena F et al (2002) Preliminary results of an early clinical experience with the Acrobot™ system for total knee replacement surgery. In: Medical image computing and computer-assisted intervention (MICCAI). Springer, London, pp 256–263

8. Taylor R, Jensen P, Whitcomb L et al (1999) A steady-hand robotic system for microsurgical augmentation. Int J Robot Res 18:1201–1210

9. Kapoor A, Kumar R, Taylor RH (2003) Simple biomanipulation tasks with "steady hand" cooperative manipulator. In: Medical image computing and computer-assisted intervention (MICCAI), pp 141–148

10. Payne CJ, Yang GZ (2014) Hand-held medical robots. Ann Biomed Eng 42:1594–1605

11. Elble RJ, Koller WC (1990) The physiology of normal tremor. In: Tremor. The Johns Hopkins University Press, Baltimore, MD

12. MacLachlan RA, Becker BC, Cuevas Tabares J et al (2012) Micron: an actively stabilized handheld tool for microsurgery. IEEE Trans Robot 28:195–212

13. Latt WT, Tan U, Shee CY, et al (2009) A compact hand-held active physiological tremor compensation instrument. IEEE/ASME Int Conf Adv Intell Mechatronics 711–716

14. Saxena A, Patel R V. (2013) An active handheld device for compensation of physiological tremor using an ionic polymer metallic composite actuator. IEEE Int Conf Intell Robot Syst 4275–4280

15. Chang D, Gu GM, Kim J (2013) Design of a novel tremor suppression device using a linear delta manipulator for micromanipulation. IEEE/RSJ Int Conf Intell Robot Syst 413–418

16. Song C, Gehlbach PL, Kang JU (2012) Active tremor cancellation by a "Smart" handheld vitreoretinal microsurgical tool using swept source optical coherence tomography. Opt Express 20:3315–3317

17. Riviere CN, Ang WT, Khosla PK (2003) Toward active tremor canceling in handheld microsurgical instruments. IEEE Trans Robot Autom 19:793–800

18. Yang S, MacLachlan RA, Riviere CN (2015) Manipulator design and operation of a six-degree-of-freedom handheld tremor-canceling microsurgical instrument. IEEE/ASME Trans Mechatron 20:761–772

19. MacLachlan RA, Riviere CN (2008) High-speed microscale optical tracking using digital fequency-domain multiplexing. IEEE Trans Instrum Meas 58(6):1991–2001

20. Yang S, Martel JN, Lobes LA et al (2018) Techniques for robot-aided intraocular surgery using monocular vision. Int J Robot Res 37:931–952

21. Becker BC, Voros S, MacLachlan RA, et al (2009) Active guidance of a handheld micromanipulator using visual servoing. IEEE Int Conf Robot Autom 339–344

22. Becker BC, MacLachlan RA, Hager GD et al (2011) Handheld micromanipulation with vision-based virtual fixtures. IEEE Int Conf Robot Autom 4127–4132

23. Becker BC, Voros S, Lobes LA et al (2010) Retinal vessel cannulation with an image-guided handheld robot. IEEE Eng Med Biol Soc Conf 5420–5423

24. Becker BC, MacLachlan RA, Lobes LA et al (2010) Semiautomated intraocular laser surgery using handheld instruments. Lasers Surg Med 42:264–273

25. Yang S, Lobes LA, Martel JN et al (2015) Hand-held automated microsurgical instrumentation for intraocular laser surgery. Lasers Surg Med 47:658–668

26. Stetten G, Wu B, Klatzky R, et al (2011) Hand-held force magnifier for surgical instruments. In: Medical image computing and computer-assisted intervention (MICCAI), pp 90–100

27. Lee R, Wu B, Klatzky R et al (2013) Hand-held force magnifier for surgical instruments: evolution toward a clinical device. Lect Notes Comput Sci (including Subser Lect Notes Artif Intell Lect Notes Bioinformatics) 7815:77–89

28. Payne CJ, Latt WT, Yang G (2012) A new hand-held force-amplifying device for micromanipulation. IEEE Int Conf Robot Autom 2012:1583–1588

29. Payne CJ (2015) Ungrounded haptic-feedback for hand-held surgical robots. https://core.ac.uk/download/pdf/77003165.pdf

30. Payne CJ, Marcus HJ, Yang GZ (2015) A smart haptic hand-held device for neurosurgical microdissection. Ann Biomed Eng 43:2185–2195

31. Marcus HJ, Payne CJ, Kailaya-Vasa A et al (2016) A "Smart" force-limiting instrument for microsurgery: laboratory and in vivo validation. PLoS One 11:1–9

32. Payne CJ, Rafii-Tari H, Marcus HJ et al (2014) Hand-held microsurgical forceps with force-feedback for micromanipulation. Proc IEEE Int Conf Robot Autom 284–289

33. Watanabe T, Koyama T, Yoneyama T et al (2017) A force-visualized silicone retractor attachable to surgical suction pipes. Sensors 17:1–18

34. Kane G, Eggers G, Boesecke R, et al (2009) System design of a hand-held mobile robot for craniotomy. In: Medical image computing and computer-assisted intervention (MICCAI), pp 402–409

35. Moccia S, Foti S, Routray A et al (2018) Toward improving safety in neurosurgery with an active handheld instrument. Ann Biomed Eng 46:1450–1464

36. Gras G, Marcus HJ, Payne CJ, et al (2015) Visual force feedback for hand-held microsurgical instruments. In: MICCAI, pp 480–487

37. Payne CJ, Kwok K, Yang G (2014) An ungrounded hand-held surgical device incorporating active constraints with force-feedback. In: IEEE/RSJ International Conference on Intelligent robots and systems (IROS), pp 2559–2565

38. Jabbour JM, Saldua MA, Bixler JN et al (2012) Confocal endomicroscopy: instrumentation and medical applications. Ann Biomed Eng 40:378–397

39. Sankar T, Delaney PM, Ryan RW et al (2010) Miniaturized handheld confocal microscopy for neurosurgery: results in an experimental glioblastoma model. Neurosurgery 66:410–418

40. Foersch S, Heimann A, Ayyad A et al (2012) Confocal laser endomicroscopy for diagnosis and histomorphologic imaging of brain tumors in vivo. PLoS One 7:e41760

41. Eschbacher J, Martirosyan NL, Nakaji P et al (2012) In vivo intraoperative confocal microscopy for real-time histopathological imaging of brain tumors. J Neurosurg 116:854–860

42. Zehri AH, Ramey W, Georges JF et al (2014) Neurosurgical confocal endomicroscopy: a review of contrast agents, confocal systems, and future imaging modalities. Surg Neurol Int 5:60

43. Latt WT, Newton RC, Visentini-Scarzanella M et al (2011) A hand-held instrument to maintain steady tissue contact during probe-based confocal laser endomicroscopy. IEEE Trans Biomed Eng 58:2694–2703

44. Giataganas P, Hughes M, Payne CJ et al (2019) Intraoperative robotic-assisted large-area high-speed microscopic imaging and intervention. IEEE Trans Biomed Eng 66:208–216

45. Hughes M, Yang G-Z (2016) Line-scanning fiber bundle endomicroscopy with a virtual detector slit. Biomed Opt Express 7:2257

46. Wisanuvej P, Giataganas P, Leibrandt K et al (2017) Three-dimensional robotic-assisted endomicroscopy with a force adaptive robotic arm. In: IEEE International Conference on Robotics and automation (ICRA). IEEE, New York, pp 2379–2384

Chapter 5

Translation

Hani J. Marcus and Christopher J. Payne

Abstract

Translation is the process by which medical innovations, such as neurosurgical robots, are brought from the laboratory into the operating room. This process has been described as a continuum punctuated by several well-defined chasms: first, the development of a robot culminating in a first-in-human study; second, the evaluation of a robot resulting in regulatory clearance or approval; and third, the adoption of a robot by neurosurgeons. In this chapter we discuss the process of innovation, and the stages of translation, the central tenet being that development and evaluation can and should proceed together in an ordered and logical manner. We also provide some practical advice on how innovators can optimize their likelihood of success.

Key words Translation, Robotics, Neurosurgery, First-in-human, Clinical studies

1 Introduction

Translation is the process by which neurosurgical robots are brought from the laboratory into the operating room. This process has been described as a continuum punctuated by several well-defined chasms: first, the development of a robot culminating in a first-in-human study; second, the evaluation of a robot resulting in regulatory clearance or approval; and third, the adoption of a robot by neurosurgeons. It is also critical for innovators to understand that translation is also a process of commercialization.

The translation of robots and other devices can take many years, and many such devices, particularly if developed in academia, are never translated. On the other hand, robots that are cleared for use are often introduced into the operating room in an unstructured and variable fashion, with obvious risks. The Balliol collaboration has proposed the IDEAL (Idea, Development, Evaluation, Assessment, Long-term study) model for device innovation to address these conflicting targets, the central tenet being that development and evaluation can and should proceed together in an ordered and logical manner (Table 1) [1–4].

Hani J. Marcus and Christopher J. Payne (eds.), *Neurosurgical Robotics*, Neuromethods, vol. 162,
https://doi.org/10.1007/978-1-0716-0993-4_5, © Springer Science+Business Media, LLC, part of Springer Nature 2021

Table 1
IDEAL stages and their characteristics

Stage	Purpose	Patients	Surgeons	Studies
Idea	Proof of concept	Very few and highly selected	Very few	Structured case reports
Development	Development	Few and selected	Few	Prospective development studies
Exploration	Learning	Many	Many	Research databases and feasibility randomized controlled studies
Assessment	Assessment	Many and expanded indications	Many	Randomized controlled studies
Long-term monitoring	Surveillance	All eligible	All eligible	Prospective registries

In this chapter we discuss the process of innovation and the stages of translation. We also provide some practical advice on how innovators can optimize their likelihood of success.

2   Needs-Driven Innovation

Robotics is an emerging technology that has significant potential to transform neurosurgery and displace existing approaches. Nonetheless, in medical device translation it is usually critical to start with a clinical need before looking to new technology. The Biodesign Innovation Process is a well-established medical device translation methodology developed at Stanford University which advocates a development process driven by the clinical need [5]. Within this framework, innovators are required to start by formulating a clinical problem from first-hand observations and close involvement with healthcare professionals (*see* Fig. 1). The Biodesign process advocates that clinical needs are formulated into broad statements that encapsulate the clinical problem, the target clinical population, and a measurable performance outcome. These problem statements should be formulated in a broad manner without reference to any specific solution. By formulating a clinical problem in this way, a complete range of solutions can be considered and critically evaluated so that an optimum solution can be selected. This point is particularly pertinent to advocates of robotics who should consider both robotic and nonrobotic solutions to unmet needs in neurosurgery. While a clear, compelling clinical need is a critical starting point for translation, it is not the only condition that is required for translation success. Within the Biodesign process, a needs-screening phase is performed to consider many other factors that will lead to the most compelling clinical needs being addressed. For example, a stakeholder analysis that considers not only the clinician

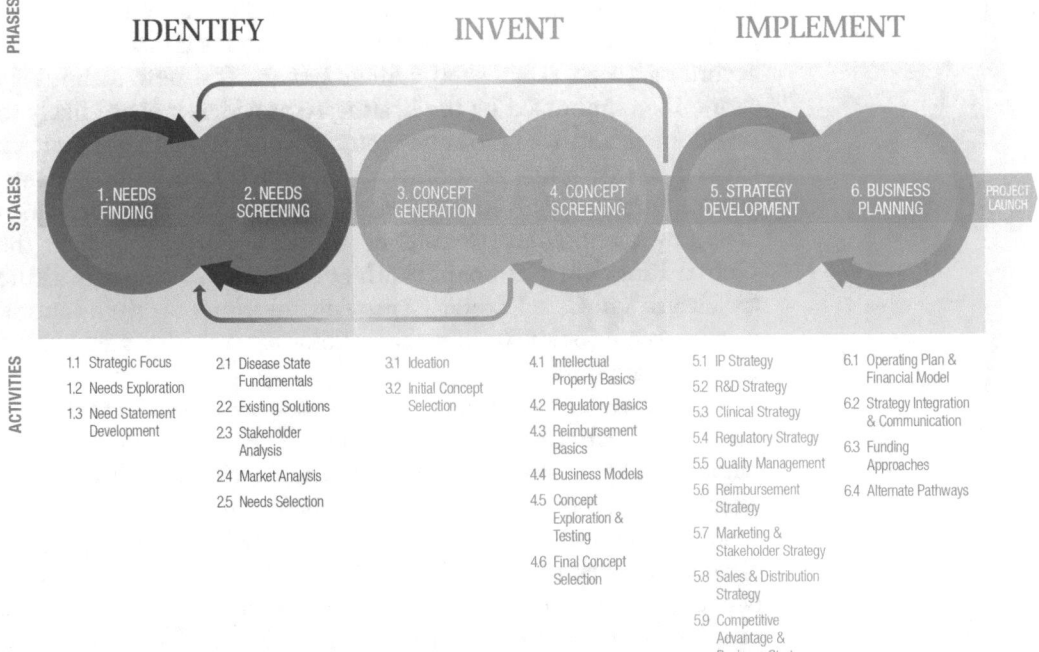

Fig. 1 The Stanford Biodesign Innovation Process for medical device translation. (Reproduced from *Biodesign: The Process of Innovating Medical Technologies*©, Second Edition, Cambridge University Press, 2015 by Paul G. Yock et al. This image is subject to copyright. Reproduced with permission of The Licensor through PLSclear)

or the patient, but also the healthcare provider, regulatory body and funders, will help guide if an identified clinical problem should be pursued or not. Critically, when screening clinical needs, it is paramount to evaluate the potential commercial value that solving the need would yield.

3   New Concept Development

Once a compelling clinical need has been found, innovators can brainstorm a wide range of potential solutions. In the same way that clinical needs are screened, device concepts must also be screened from the perspective of all stakeholders; and critically evaluated against commercial viability criteria. There are many factors that determine the commercial viability of a concept solution; these include the market size, product economics, intellectual property landscape, revenue generation, funding considerations, competition (including alternative approaches to the same need), regulatory strategy and clinical strategy.

When screening design concepts, innovators must ensure there is a clear mechanism by which healthcare providers can purchase new technologies. In the United States, medical devices purchased by the healthcare provider are reimbursed by insurance companies according to an established coding system. If a new technology cannot be reimbursed by the healthcare provider, it is not likely to be adopted and so innovators must ensure a billing mechanism exists for emerging technologies. In public healthcare systems, governmental organizations advise healthcare providers on which emerging technologies should be adopted. For example, in the United Kingdom National Health Service, the National Institute for Health and Care Excellence provide guidelines for the adoption of new technologies based on safety, efficacy and cost effectiveness of new treatments.

The commercial viability of a potential solution must consider the product economics at an early stage. The translation of robotic platforms require significant research and development investment which must be recouped. A key stakeholder in this respect is the funder, commonly venture capitalists, who will only fund projects capable of generating significant returns on investment. For this to happen, a market opportunity must be sufficiently large, and the proposed solution must be priced sufficiently high. This latter point introduces another key stakeholder: the healthcare provider. To justify the costs of adopting a new medical device such as a robotic system, the healthcare provider must also reap economic value. In this respect, Egeland et al. make a curious observation about healthcare economics: it is typically the healthcare provider who pays for a surgical device but the value of the device might primarily benefit the patient [6]. Whilst it is clearly important that new neurosurgical robotic systems should be patient-focused, the interests of the healthcare provider must also be considered carefully. Egeland et al. make the point that it is the financial administrators within a healthcare provider who ultimately make purchasing decisions and control the adoption of new technologies. Therefore, innovators should consider developing technologies that provide direct economic value to the healthcare provider. Egeland et al. highlight three ways in which new surgical technologies can be economically beneficial to healthcare providers.

Firstly, new surgical devices can provide economical value to a healthcare provider by improving efficiency. If a new robotic technology can reduce procedure or turnover times, this can allow for a greater patient throughput which can yield significant savings. Robotic systems may allow neurosurgeons to perform complex procedures faster, but these gains must be offset against the robot setup time. In this respect, the clinical workflow should be a considered carefully in the concept design phase to ensure operating times can be reduced. The second means by which a robotic system may provide economic value to the healthcare provider is by preventing high costs. If a robotic system can prevent or limit postoperative complications, the healthcare provider can make significant savings. For example, if a robotic system can reduce the invasiveness of a procedure to tangibly reduce postoperative recovery times and overall hospital stay, this directly translates into significant cost savings. Robotics may also allow less experienced clinicians to perform procedures to the same standard as experienced clinicians which can also yield cost benefits to the healthcare provider. Thirdly, if a new surgical device can undercut a more expensive status quo, the healthcare provider will make a direct saving. This final category is likely to be particularly challenging for robotic technology which is inherently expensive to develop and manufacture.

When considering design concepts, innovators must perform cost-effectiveness analyses that predict how a new solution will create economic value. During the concept development process, device solutions should be evaluated critically against these outlined commercial realities. The process is iterative and eventually an optimal solution that satisfies all stakeholders and commercial viability criteria will emerge as the most compelling opportunity to pursue.

4   Intellectual Property and Competitive Analysis

Intellectual property (IP) is a broad term that refers to a wide range of creations of the mind [7]. IP can be protected by law through patents, copyright, and trademarks. In the concept development process, it is essential to review the intellectual property surrounding existing solutions to the clinical need. New device innovations are likely to require some form of intellectual property that will act as a barrier-to-entry for competitor products. Furthermore, it is important that any new device concept be distinct from prior art for a patent to be successfully granted.

It is usually important to obtain patents that are broad in scope to cover all potential variations of an innovation, increasing the value and level of protection against competitors. The field of intellectual property related to medical devices and surgical robotics is colossal, and so it will be critical for innovators to perform rigorous prior art searches early on to avoid or minimize litigation challenges at a later stage in translation; and to increase the fundability of a project.

Aside from IP, innovators will also need to perform an analysis of current and emerging competitors to provide a comprehensive understanding of existing products; and how their proposed solution will offer a performance advantage. Furthermore, innovators need to carefully consider both direct and indirect competitors in evaluating potential solutions. In neurosurgery, direct competitors might include existing robotic platforms or surgical tools, but any other approaches to the same clinical problem such as therapeutic drugs or radiological treatments must also be considered.

5   Preclinical Studies, Design Controls, and First-In-Human Study

The development of a robot culminating in a first-in-human study is arguably the most challenging stage of translation. At this stage, innovators will have begun development of a solution that meets a compelling clinical and market need. Once it has been determined that the proposed solution has the potential to be commercialized, development can proceed toward a first-in-human study. Currently, there is no consensus on what studies are required before embarking on a first-in-human study, but we would recommend several stakeholder perspectives be considered (Table 2). We have previously discussed the perspectives of healthcare provider and funder in evaluating the commercial viability of a medical device solution. Additionally, a patient must consider the robot *acceptable*, which may be assessed through patient surveys and focus groups. The neurosurgeon end-users must consider the robot *clinically usable* which may be assessed through clinician surveys and focus groups. Furthermore, the device must be both *safe* and *effective*, which may be assessed through laboratory, cadaver, and animal studies.

Like all medical devices, robotic systems under development must be subject to *design controls*. Design controls are a formalized set of practices and procedures within the design development process that seek to ensure a developed device is appropriate for its intended use (Fig. 2). The United States Food and Drug Administration (FDA) *1997 Design control guidance for medical device manufacturers* is an excellent resource that outlines the process in detail [8]. Implementation of design controls provides designers and managers greater visibility of the design process and allows errors to be spotted, and corrected, early on in the

Table 2
Preclinical studies classified according to perspective

Perspective	Preclinical studies
Funder	Economic analyses Unmet needs analyses
Healthcare institution	Economic analyses Unmet needs analyses
Patient	Patient surveys Patient focus groups
Neurosurgeon	Clinician surveys Clinician focus groups
Robot	Laboratory studies Cadaver studies Animal studies

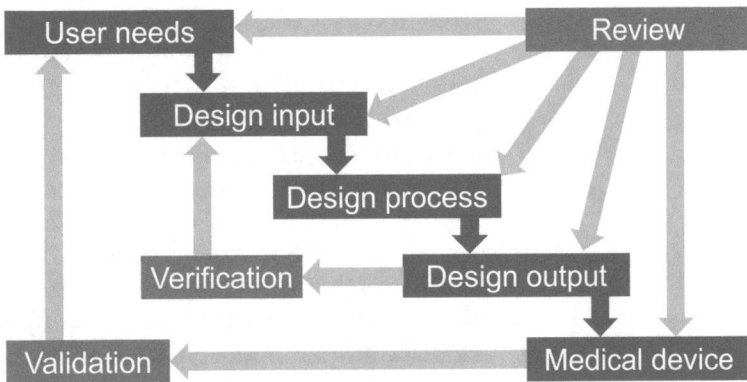

Fig. 2 Design controls waterfall diagram, as outlined in [8]

development process. The process starts by gathering the needs of the medical device of the new medical device. Needs are converted into design input requirements which are a series of quantifiable performance characteristics. For each design input, there is a design output which are the results of the design effort. Design verification is the process that confirms an individual design output meets the design input through objective evidence. Design validation is the process that evaluates that the developed medical device meets the intended user needs. Throughout this process, regular design reviews, conducted at strategic points in the development process, ensure that all requirements will be met, and that discrepancies are identified and corrected.

Once a robotic system has been developed and validated, a first-in-human study can be performed. The IDEAL recommendations (Stage 1) are that this first-in-human study is registered in a clinical trials database and that the findings are published as a structured case report, ideally in a peer-reviewed journal.

Sadly, even the most promising robots, published in journals with the highest impact factors, are rarely translated. In a bibliometric study of all new surgical devices reported in the biomedical engineering literature between January 1993 and January 2000, we found that only one in ten resulted in a first-in-human study within a decade of publication [9]. Importantly, early collaboration between clinicians and engineers was an important predictor of success, and devices developed with early collaboration were six times more likely to be translated than those without.

6   Further Clinical Studies and Regulatory Approval

The evaluation of a robot in clinical trials culminating in regulatory clearance or approval is very inconsistent. Indeed, most devices are cleared for use before peer-reviewed clinical trials are published [10].

The IDEAL recommendations (Stage 2) are that robots and other surgical devices are evaluated in small uncontrolled cohort studies, usually within a single center, with consecutive case reporting and explanation of timing and rationale for changes in procedure [4]. This is usually followed by a large uncontrolled prospective cohort study, evaluating learning curves and building consensus on quality control. In order to minimize risk to patients, we would suggest these studies be performed in experienced centers.

Where a robot-assisted procedure is gaining wide acceptance and considered as a possible replacement for the standard approach, the IDEAL recommendation (Stage 3) is that a randomized controlled study be considered to provide a definitive clinical comparison [4]. If such a study is performed, we would suggest doing so after learning curves are overcome to avoid any confounding effects.

The FDA recognizes several types of regulatory pathway depending on the nature of the device. If a device is "substantially equivalent" to a predicate device then premarket notification 510 (k) is appropriate, which does not necessarily require any clinical data. If, however, a device is "not substantially equivalent" then premarket approval (PMA) is appropriate, and a reasonable evidence of safety and effectiveness is required. There are also other regulatory pathways such as humanitarian device exemption (HDE) if the device is for use in patients with rare diseases, but these are rarely used within surgical robotics.

The European Union and other regulatory bodies have analogous pathways but usually require less evidence of safety and effectiveness before market approval.

Irrespective of which pathway is used, securing regulatory clearance or approval is challenging. In a bibliometric study of new surgical devices reported in the academic literature, 99 (45%), ultimately received regulatory clearance or approval including 510(k) clearance (78/99; 79%), PMA (17/99; 17%), and HDE (4/99, 4%) [10]. The most important predictor of success was industry collaboration, and devices developed with collaboration were four times more likely to be translated than those without.

7 Dissemination

The dissemination of robots into widespread practice is perhaps the least understood stage of translation.

The IDEAL recommendation (Stage 4) is that once a robot or other surgical device has been approved, its use should be monitored in registries to identify any late or rare problems, and changes in use. Such registries are particularly important for "me-too" robots that enter surgical practice through the 510(k) pathway.

Table 3
Factors predicting successful dissemination according to perspective

Perspective	Factors
Robot	Relative advantage
	Clinical benefits clearly visible
	Easy to trial before purchasing
	Easy to learn to use
	Compatible with existing hardware
	Potential for use in other indications
Neurosurgeon	Highly able
	Highly motivated
Healthcare institution	Unmet need
	Financial benefits clearly visible
	Fits within values and norms

Perhaps the best model to understand the dissemination of robots into widespread practice is the Diffusion of Innovations theory by Everett Rogers [11]. Although the theory was initially used to examine the how farmers adopted advances such as hybrid seeds in the Midwestern United States in the 1920s and 1930s, it has been applied to numerous contexts including robotics and neurosurgery [12]. When predicting whether a robot will be successfully disseminated, we would recommend considering the characteristics of the robot, the neurosurgeon, and the healthcare system (Table 3).

There are several features of the robot that may facilitate dissemination. Most obviously, a robot with a relative advantage over standard surgery is more likely to be disseminated than a robot without, and this purported advantage may be borne out of the clinical studies and regulatory approval process. A related but distinct point is that if this benefit is clearly observable then it will result in greater dissemination, for example, allowing a surgeon to perform minimal access surgery when they might otherwise resort to an open approach.

Another feature of the robot that may favor rapid dissemination is compatibility with existing surgical devices and workflow, for example, interfacing with existing image guidance platforms or endoscopes. Similarly, a robot that appears familiar and is simple to use may also be more easily disseminated.

A robot that can be easily trialed by a neurosurgeon or healthcare institution may also hold an advantage over a robot that is not. This is particularly pertinent in the context of historically large robotic platforms that are rather cumbersome to place in the operating room for a short trial period.

Often, a robot designed for a particular application may find use in other related applications and this can also allow for more rapid dissemination. A case in point is a supervisory-control robot that can be used for trajectory planning in pedicle screws placement, deep brain stimulation, or brain biopsy.

Others features of a robot that have been suggested as influencing dissemination include being less risky and less disruptive, though these are contentious.

Alongside these characteristics of the robot, it is important to consider those of the neurosurgeon. It is generally considered that a body of neurosurgeons with greater ability and motivation will be more likely to adopt innovations such as robotics, and this may vary between subspecialties.

Finally, there are several characteristics of a healthcare system that may allow for more rapid adoption. These include an identified unmet need that the robot fulfils, and the long-term financial benefits of the robot being clearly visible. Less tangible but, nonetheless, important, a healthcare system with values and norms that complement a robot's use will be more likely to adopt said robot.

8   Conclusions

The translation of a robot from the laboratory into the operating room is challenging. However, there are several important points to consider that may facilitate translation, particularly for a robot developed in academia. Needs-driven innovation is a critical starting point for successful translation. Early collaboration between engineers and clinicians will more likely result in a first-in-human study. It is also vitally important for innovators to garner a firm understanding of the commercial realities of the clinical problem at an early stage in the innovation process. Early stakeholder analyses will help innovators select the correct clinical problems to address and identify the optimal solutions to pursue. Competitor and IP analyses are also import facets to consider in the earlier stages of development. For academic innovators, industry collaboration will more likely result in regulatory approval or clearance; and a structured introduction of the robot into clinical practice will allow for both improved safety and demonstration of effectiveness.

References

1. Cook JA, McCulloch P, Blazeby JM, Beard DJ, Marinac-Dabic D, Sedrakyan A, Grp I (2013) IDEAL framework for surgical innovation 3: randomised controlled trials in the assessment stage and evaluations in the long term study stage. BMJ 346:Artn F2820. https://doi.org/10.1136/Bmj.F2820

2. Ergina PL, Barkun JS, McCulloch P, Cook JA, Altman DG, Grp I (2013) IDEAL framework for surgical innovation 2: observational studies in the exploration and assessment stages. BMJ 346:Artn F3011. https://doi.org/10.1136/Bmj.F3011

3. McCulloch P, Cook JA, Altman DG, Heneghan C, Diener MK, Grp I (2013) IDEAL framework for surgical innovation 1: the idea and development stages. BMJ 346:

Artn F3012. https://doi.org/10.1136/Bmj. F3012

4. Sedrakyan A, Campbell B, Merino JG, Kuntz R, Hirst A, McCulloch P (2016) IDEAL-D: a rational framework for evaluating and regulating the use of medical devices. BMJ 353:i2372. https://doi.org/10.1136/bmj. i2372

5. Biodesign: the process of innovating medical technologies, 2nd edn

6. Egeland RD, Rapp Z, David FS (2017) From innovation to market adoption in the operating room: the "CFO as customer". Surgery 162 (3):477–482. https://doi.org/10.1016/j. surg.2017.04.007

7. WIPO (2020) What is intellectual property? https://www.wipo.int/about-ip/en/

8. FDA (1997) Design control guidance for medical device manufacturers

9. Marcus HJ, Payne CJ, Hughes-Hallett A, Gras G, Leibrandt K, Nandi D, Yang GZ (2016) Making the leap: the translation of innovative surgical devices from the laboratory to the operating room. Ann Surg 263 (6):1077–1078. https://doi.org/10.1097/ SLA.0000000000001532

10. Marcus HJ, Payne CJ, Hughes-Hallett A, Marcus AP, Yang GZ, Darzi A, Nandi D (2016) Regulatory approval of new medical devices: cross sectional study. BMJ 353:i2587. https://doi.org/10.1136/bmj.i2587

11. Rogers E (1962) Diffusion of innovations, 5th edn. Free Press, New York

12. Marcus HJ, Hughes-Hallett A, Kwasnicki RM, Darzi A, Yang GZ, Nandi D (2015) Technological innovation in neurosurgery: a quantitative study. J Neurosurg 123:174–181. https:// doi.org/10.3171/2014.12.JNS141422

Part II

Application

Chapter 6

Robot-Assisted Brain Biopsy

Michel Lefranc

Abstract

Stereotactic biopsy is a routine neurosurgical procedure. Although a robot was first used to perform a stereotactic biopsy in 1985, little on robot-assisted stereotactic biopsy has been published since then. However, robot-assisted stereotactic biopsy appears to be a safe and effective way of establishing a histological diagnosis that combines the advantages of frame-based and frameless techniques. The superiority of robot-assisted stereotactic biopsy over standard surgical techniques is yet to be established.

Key words Robotic assistance, Robot, Stereotactic biopsy, Brain tumor

1 Introduction

Stereotactic biopsy is a routine neurosurgical procedure, the purpose of which is to obtain an accurate histological diagnosis with the least possible morbidity. In the 1970s, cranial biopsies were performed using a free-hand technique. Surgeons moved to frame-based stereotactic biopsies in the 1980s and 1990s because of their demonstrated safety and accuracy [1, 2]. Nowadays, frameless techniques have been adopted by most neurosurgeons. However, frame-based techniques are still indicated in specific settings [2, 3].

The first reported robot-assisted stereotactic brain biopsy was performed in 1985, when a modified industrial robot was used to define the biopsy's trajectory [4]. Indeed, this biopsy constituted the first ever use of robotic assistance in neurosurgery.

Since this first report, many types of surgical robot (including the neuromate® (Renishaw, Gloucestershire, UK), the ROSA® (Medtech, Montpellier, France), the iSYS 1 (iSYS, Kitzbühel, Austria), and the Renaissance® (Mazor Robotics, Caesarea, Israel) have been used to perform stereotactic brain biopsies and other stereotactic procedures such as the placement of deep brain stimulation (DBS) electrodes, and stereo-electro-encephalography (SEEG) [5].

Hani J. Marcus and Christopher J. Payne (eds.), *Neurosurgical Robotics*, Neuromethods, vol. 162,
https://doi.org/10.1007/978-1-0716-0993-4_6, © Springer Science+Business Media, LLC, part of Springer Nature 2021

For a brain biopsy, the objective of robotic assistance is to combine the accuracy of frame-based stereotactic technique with the facilitated workflow and shorter operating time associated with frameless biopsies. The clinical corollary is a safer, more effective surgical intervention.

2   Summary of Evidence

Although little data is available, it appears that robot-assisted stereotactic brain surgery is starting to be adopted by the neurosurgical community—especially for complex stereotactic procedures such as SEEG, laser ablation and DBS surgery [6–11].

Most of the neurosurgical literature on robot-assisted biopsy is based on retrospective cases series; these have evidenced a high diagnostic accuracy (>95%) and a low morbidity rate [12–19].

The level of evidence is also limited by the great heterogeneity of robotic devices and registration modes [16, 20]. All the robots described in the literature have a supervisory-control function in which the surgeon planned the surgical trajectory (using the pre-operative volumetric 3D imaging data) and the robot then carries out trajectory alignment autonomously under the supervision of the surgeon [21]. However, the rest of the surgical technique varies widely with regard to (a) the images used for the registration process (MRI or CT), (b) the image resolution, and (c) the registration process using frame-based or frameless methods, (one or more bone fiducials, or surface registration using MRI or CT data).

Image-guided stereotactic biopsy is obviously influenced by the image quality and the registration process [22]. Thus, the biopsy's overall accuracy is impacted by every steps and technical choice during the surgical process more than by use of the robotic device itself. This may explain the differences in accuracy reported in the literature.

When average target accuracy was reported, it ranges from 0.9 to 4.5 mm [16]. Unfortunately, most publications do not specify the registration method. Willems et al. compared the target accuracy of the MKM robot (Zeiss, Oberkochen, Germany) using different registration methods, and found that bone-anchored fiducials resulted in significantly greater accuracy than adhesive scalp markers (3.3 ± 1.7 versus 4.5 ± 2 mm, respectively) [23]. Our department evaluated the accuracy of the ROSA® robot; similarly, we found that the accuracy with frame-based registration (0.81 ± 0.39 mm) and bone-fiducial registration (0.7 ± 0.42 mm) was greater than that with surface registration (1.22 ± 0.73 mm). Use of MRI as the reference image was associated with significantly lower overall accuracy—probably due to distortion in the MRI system [22].

Only two studies have provided separate accuracy data for the entry and target points. Minchev et al. reported on use of the iSYS 1 robot in 25 patients; the median (range) entry point accuracy was 1.3 mm (0.2–2.6 mm), and the median target point accuracy was 0.9 mm (0.0–3.1 mm) [17]. Dlaka et al. reported on the RONNA G3 robot in a single patient; the entry point error was 2.2 mm, and target point error was 2.3 mm [24].

In contrast to frame evaluations [25], the mechanical accuracy of the use of these robotic devices in humans has not been evaluated. Significant differences are likely to be found, given the major design difference between devices (tiny arms on the frame for devices such as the iSYS 1 and Renaissance®, versus an industrial robotic arm on a 200 kg trolley that also supports the patient's head [16, 20]).

It must be noted that in the absence of comparative trials of frameless robot-assisted biopsy approaches vs. conventional frame-based approaches, it is not known which technique is superior. However, the literature data shows that robot-assisted stereotactic biopsy is at least as effective and safe as existing frame-based and frameless biopsy techniques.

Marcus et al. recently reviewed the literature on robot-assisted biopsy [16]. Six studies reported on the use of the ROSA® robot, and two reported on the neuromate® robot; the other robots used were the Puma 200, the Renaissance®, the Minerva (Swiss Federal Institute of Technology, Lausanne, Switzerland), the MKM, the Surgiscope (ISIS, Grenoble, France), the iSYS 1, and the RONNA G3 (University of Zagreb, Zagreb, Croatia).

In individual series, the diagnostic rate for robot-assisted biopsy varied from 75% to 100%. When the data were pooled, the diagnostic biopsy rate was 94.9% [16]. The diagnostic accuracy of robotic stereotactic biopsy appears to be at least as high as that of conventional frame-based and frameless stereotactic procedures [12–19].

The overall morbidity rate appears to be very low, and is essentially related to haemorrhage. In Marcus et al.'s literature review of pooled data from all the robot-assisted studies, it was found that post-operative haematoma was reported in 7.5% of cases. Symptomatic haemorrhage that required craniotomy and evacuation of the haematoma occurred in 0.7% cases. The incidence of neurological deficits was 5.1%, and that of permanent neurological deficits was below 1%. No mortalities were reported in the pooled analysis [16].

Advantages in terms of ease of use, operating time or workflow are often mentioned but rarely quantified [15, 16]. In our experience, shorter operating times and lower operating room occupation are major advantages of robot-assisted biopsy.

Lastly, further evaluation (according to the IDEAL guidelines, with research databases and comparative trials) of the clinical advantages of robot-assisted stereotactic biopsies is required to assess the superiority of robot assisted techniques.

3 Description of Robot

At our center, we use the ROSA® robot. The robot has a supervisory-control mode in which the surgeon plans the surgical trajectory on the basis of pre-operative volumetric 3D imaging. After robot-assisted registration, the robotic arm is then moved into place in order to align the instruments with the planned trajectory. The surgeon then performs the surgery (drilling, coagulation, and biopsy) through a working channel held by the robot; all the procedures are performed along the planned trajectory (Fig. 1).

The ROSA robot has a mechanical arm with six degrees of freedom. The robot has a large footprint, and weighs up to 200 kg.

Registration to the patient is either with laser scanning of anatomic landmarks (facial features) or by way of touch alignment to surgically implanted in-bone skull fiducials.

4 Description of the Technique

4.1 Stereotactic Planning

Biopsy needle targeting and trajectory determination are performed using the robot's planning software. The target and trajectory are planned before surgery. Targeting is usually based on MR images. The surgical planning determines a gyral entry point, in order to avoid a trans-sulcus trajectory or stripping of the ependymal wall. Target points are situated over a region with hyperperfusion, contrast enhancement or, in the absence of either of these two signs, a hyperintense signal in FLAIR or T2-weighted sequences [15].

4.2 Surgical Procedure

Here, we describe the procedures for the Rosa® robot device in frameless mode; the author primarily uses this device and this configuration for stereotactic biopsies.

Although the robot-assisted surgical procedures are generally very similar, the registration method, head fixation and fiducial design can differ significantly from one robot to another. Here, we describe surface registration using either anatomic landmarks or bone fiducials.

Most procedures are performed with the patient under general anaesthesia. For a biopsy using anatomic landmarks, the robot is fitted with a MAYFIELD® headrest (Integra LifeSciences Corporation). The head is therefore securely attached to the robotic device via a rigid arm, in order to avoid any mechanical movement.

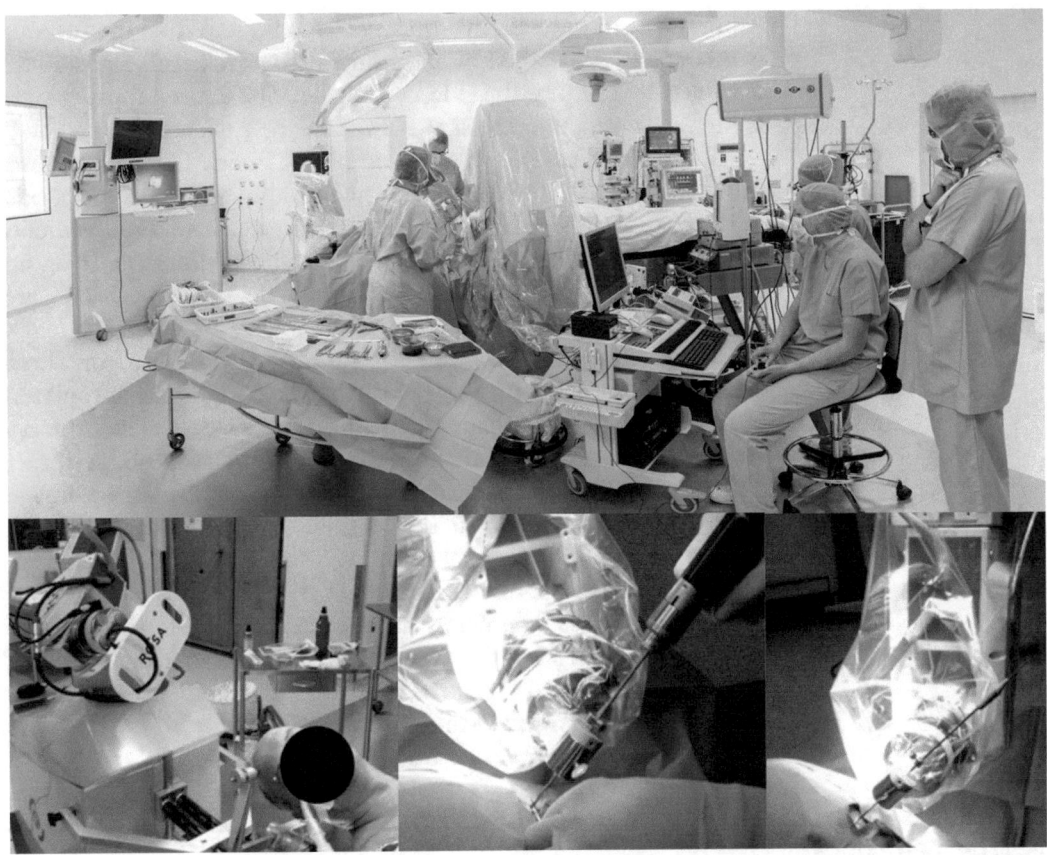

Fig. 1 Example of operating room setup (top), position of robot with respect to patient (bottom left), and use of robot to define trajectory of drilling and biopsy (bottom centre and bottom right respectively)

Automatic robotic frameless surface registration is then performed using laser scanning of anatomical landmarks (facial features). The accuracy of registration is confirmed by the surgeon with reference to several landmarks, such as the root of the nose, internal and external canthi, temples, midline, and any free landmarks chosen by the surgeon. The operating field is draped, and the robotic arm is automatically positioned along the planned trajectory. The robot is used as an instrument holder during stereotactic surgery (as arcs with frames). Thus, the surgeon passes instruments through the reducers. All instruments are then positioned and used via an appropriate reducer, held by the robotic arm. A 3.2-mm hole is drilled along the trajectory. The dura is opened by electrocautery with a bushing-guided blunt stylet that is insulated except at its tip. A Sedan side-cut biopsy needle (specimen window: 10 mm; diameter: 2.5 mm) is aligned with the trajectory. Staged biopsies (at least two stages) are performed to obtain as much tissue as possible and to optimize sample collection. Each stage produces 4 "rosette" biopsy samples in 4 quadrants by rotating the side-cut needle, using the standard negative-pressure technique. When the tumour is too small for a staged biopsy, only 4 to 8 "rosette" biopsies are obtained by rotating the side-cut needle. The entire biopsy needle is then removed, and the skin is closed with 2–0 suture.

For a frameless biopsy using implanted bone fiducial markers, the MAYFIELD® headrest is similarly attached to the robotic device via a rigid arm, in order to avoid any mechanical movement. A cone beam CT device (CBCT, the O-arm Surgical Imaging System from Medtronic) is then installed. Three positions are registered (parked, to place the O-arm away from the patient's head and allow enough space for surgery; lateral and frontal, to centre the O-arm for imaging of the head) to guide the automated movements of the O-shaped gantry during surgery. The fiducials are placed on patient's head in the operating position. A three-dimensional FPCT scan is acquired and used as the reference image for registration. The CT images is then matched with the pre-op MR images used for surgical planning. Robot-assisted fiducial registration is then performed. The rest of the procedure is performed in the same way as a frameless surface registration biopsy.

Patients are discharged to home on postoperative day 1 or 2, if their clinical status is satisfactory. They return to hospital for a diagnostic consultation within 2 weeks of the biopsy (i.e. once the definitive histology report is obtained).

5 Tips and Tricks

The main errors in stereotactic surgery are related to imaging factors, the registration technique, and the mechanical accuracy of the device used.

High-quality imaging with low distortion is mandatory for accurate, precise biopsies. It is necessary to (a) check for the absence of head movements during MRI and stereotactic CT, and (b) confirm the accuracy of the automatic matching against landmarks. If an error has occurred, the matching must be repeated and corrected until it is perfect.

We advocate the use of a CT scan as the reference image because of its homogenous spatial resolution over the entire field of view. Thus, the accuracy is directly and solely related to the slice thickness and matrix size; the higher the resolution, the better the registration process. Magnetic resonance imaging has excellent contrast resolution but is subject to geometric distortions [22, 26]. Indeed, MRI scans are inhomogeneous in terms of spatial accuracy within the field of view due to non-linearity of the gradient, chemical shifts, and magnetic susceptibility [26]. This results in specific distortions at the periphery of the field of view and in regard of interfaces. These distortions are most troublesome at interfaces between two different structures—particularly at the air-scalp/skin interface but also at the periphery of the field of view which are the sites used for frame-based fiducials or several bone fiducials registration [22, 27]. With MRI scans, distortions within the center of field of view can be decreased particularly by using a high-resolution matrix, such as 512×512 pixel matrix. Fusion with CT scans, avoiding the periphery of the field of view and interface (skin/air), allows use of the MRI data to perform the plan of the surgery and the CT scans for robot's registration with highest level of accuracy.

To obtain the highest accuracy, we recommend an image with a 512×512 pixel matrix size and sub-millimetre slices (0.625 mm for CT and ≤ 1 mm for MRI scan).

The rigidity of the head holder must be checked carefully. Lastly, the planned trajectory must avoid vessels and ventricles, and (to minimize the risk of haemorrhage) must never pass through the midline.

6   Conclusions

Robotic surgery appears to enable accurate, safe and rapid frameless biopsy procedures by combining the respective advantages of frameless and frame-based techniques. Robotic assistance helps to improve the safety and effectiveness of surgical biopsy for histological diagnosis, with a low morbidity rate.

However, the superiority of robot-assisted biopsy techniques over other surgical techniques particularly frame-based remains to be established, as the few clinical series published to date included small numbers of patients.

References

1. Kondziolka D, Lunsford LD (1999) The role of stereotactic biopsy in the management of gliomas. J Neuro-Oncol 42:205–213. https://doi.org/10.1023/a:1006105415194

2. Smith JS, Quinones-Hinojosa A, Barbaro NM, McDermott MW (2005) Frame-based stereotactic biopsy remains an important diagnostic tool with distinct advantages over frameless stereotactic biopsy. J Neuro-Oncol 73:173–179. https://doi.org/10.1007/s11060-004-4208-3

3. Owen CM, Linskey ME (2009) Frame-based stereotaxy in a frameless era: current capabilities, relative role, and the positive- and negative predictive values of blood through the needle. J Neuro-Oncol 93:139–149. https://doi.org/10.1007/s11060-009-9871-y

4. Kwoh YS, Hou J, Jonckheere EA, Hayati S (1988) A robot with improved absolute positioning accuracy for CT guided stereotactic brain surgery. IEEE Trans Biomed Eng 35:153–160. https://doi.org/10.1109/10.1354

5. Faria C, Erlhagen W, Rito M, De Momi E, Ferrigno G, Bicho E (2015) Review of robotic technology for stereotactic neurosurgery. IEEE Rev Biomed Eng 8:125–137. https://doi.org/10.1109/RBME.2015.2428305

6. Abel TJ, Varela Osorio R, Amorim-Leite R, Mathieu F, Kahane P, Minotti L, Hoffmann D, Chabardes S (2018) Frameless robot-assisted stereoelectroencephalography in children: technical aspects and comparison with Talairach frame technique. J Neurosurg Pediatr 22:37–46. https://doi.org/10.3171/2018.1.PEDS17435

7. Gonzalez-Martinez J, Vadera S, Mullin J, Enatsu R, Alexopoulos AV, Patwardhan R,

Bingaman W, Najm I (2014) Robot-assisted stereotactic laser ablation in medically intractable epilepsy: operative technique. Neurosurgery 10(Suppl 2):163–167. https://doi.org/10.1227/NEU.0000000000000286

8. Ho AL, Pendharkar AV, Brewster R, Martinez DL, Jaffe RA, Xu LW, Miller KJ, Halpern CH (2019) Frameless robot-assisted deep brain stimulation surgery: an initial experience. Oper Neurosurg 17:424–431. https://doi.org/10.1093/ons/opy395

9. Lefranc M, Le Gars D (2012) Robotic implantation of deep brain stimulation leads, assisted by intra-operative, flat-panel CT. Acta Neurochir 154:2069–2074. https://doi.org/10.1007/s00701-012-1445-7

10. Miller BA, Salehi A, Limbrick DDJ, Smyth MD (2017) Applications of a robotic stereotactic arm for pediatric epilepsy and neurooncology surgery. J Neurosurg Pediatr 20:364–370. https://doi.org/10.3171/2017.5.PEDS1782

11. Tir M, Devos D, Blond S, Touzet G, Reyns N, Duhamel A, Cottencin O, Dujardin K, Cassim F, Destee A, Defebvre L, Krystkowiak P (2007) Exhaustive, one-year follow-up of subthalamic nucleus deep brain stimulation in a large, single-center cohort of parkinsonian patients. Neurosurgery 61:295–297. https://doi.org/10.1227/01.NEU.0000285347.50028.B9

12. Carai A, Mastronuzzi A, De Benedictis A, Messina R, Cacchione A, Miele E, Randi F, Esposito G, Trezza A, Colafati GS, Savioli A, Locatelli F, Marras CE (2017) Robot-assisted stereotactic biopsy of diffuse intrinsic pontine glioma: a single-center experience. World Neurosurg 101:584–588. https://doi.org/10.1016/j.wneu.2017.02.088

13. Coca HA, Cebula H, Benmekhbi M, Chenard MP, Entz-Werle N, Proust F (2016) Diffuse intrinsic pontine gliomas in children: interest of robotic frameless assisted biopsy. A technical note. Neurochirurgie 62:327–331. https://doi.org/10.1016/j.neuchi.2016.07.005

14. Haegelen C, Touzet G, Reyns N, Maurage C-A, Ayachi M, Blond S (2010) Stereotactic robot-guided biopsies of brain stem lesions: experience with 15 cases. Neurochirurgie 56:363–367. https://doi.org/10.1016/j.neuchi.2010.05.006

15. Lefranc M, Capel C, Pruvot-Occean A-S, Fichten A, Desenclos C, Toussaint P, Le Gars D, Peltier J (2015) Frameless robotic stereotactic biopsies: a consecutive series of 100 cases. J Neurosurg 122:342–352. https://doi.org/10.3171/2014.9.JNS14107. Disclosure

16. Marcus HJ, Vakharia VN, Ourselin S, Duncan J, Tisdall M, Aquilina K (2018) Robot-assisted stereotactic brain biopsy: systematic review and bibliometric analysis. Childs Nerv Syst 34(7):1299–1309. https://doi.org/10.1007/s00381-018-3821-y

17. Minchev G, Kronreif G, Martinez-Moreno M, Dorfer C, Micko A, Mert A, Kiesel B, Widhalm G, Knosp E, Wolfsberger S (2017) A novel miniature robotic guidance device for stereotactic neurosurgical interventions: preliminary experience with the iSYS1 robot. J Neurosurg 126:985–996. https://doi.org/10.3171/2016.1.JNS152005

18. Terrier L, Gilard V, Marguet F, Fontanilles M, Derrey S (2019) Stereotactic brain biopsy: evaluation of robot-assisted procedure in 60 patients. Acta Neurochir 161:545–552. https://doi.org/10.1007/s00701-019-03808-5

19. Yasin H, Hoff HJ, Blumcke I, Simon M (2019) Experience with 102 frameless stereotactic biopsies using the neuromate robotic device. World Neurosurg 123:e450–e456. https://doi.org/10.1016/j.wneu.2018.11.187

20. Fomenko A, Serletis D (2018) Robotic stereotaxy in cranial neurosurgery: a qualitative systematic review. Neurosurgery 83:642–650. https://doi.org/10.1093/neuros/nyx576

21. Nathoo N, Cavusoglu MC, Vogelbaum MA, Barnett GH (2005) In touch with robotics: neurosurgery for the future. Neurosurgery 56:421–433

22. Lefranc M, Capel C, Pruvot AS, Fichten A, Desenclos C, Toussaint P, Le Gars D, Peltier J (2014) The impact of the reference imaging modality, registration method and intraoperative flat-panel computed tomography on the accuracy of the ROSA(R) stereotactic robot. Stereotact Funct Neurosurg 92:242–250. https://doi.org/10.1159/000362936

23. Willems PWA, Noordmans HJ, Ramos LMP, Taphoorn MJB, Berkelbach van der Sprenkel JW, Viergever MA, Tulleken CAF (2003) Clinical evaluation of stereotactic brain biopsies with an MKM-mounted instrument holder. Acta Neurochir 145:889–897; discussion 897. https://doi.org/10.1007/s00701-003-0112-4

24. Dlaka D, Svaco M, Chudy D, Jerbic B, Sekoranja B, Suligoj F, Vidakovic J, Almahariq F, Romic D (2018) Brain biopsy performed with the RONNA G3 system: a case study on using a novel robotic navigation device for stereotactic neurosurgery. Int J Med Robot 14. https://doi.org/10.1002/rcs.1884

25. Maciunas RJ, Galloway RL Jr, Latimer JW (1994) The application accuracy of stereotactic

frames. Neurosurgery 35:682–685. https:// doi.org/10.1227/00006123-199410000-00015

26. Langlois S, Desvignes M, Constans JM, Revenu M (1999) MRI geometric distortion: a simple approach to correcting the effects of non-linear gradient fields. J Magn Reson Imaging 9:821–831. https://doi.org/10.1002/(

sici)1522-2586(199906)9:6<821::aid-jmri9>3.0.co;2-2

27. Kondziolka D, Dempsey PK, Lunsford LD, Kestle JR, Dolan EJ, Kanal E, Tasker RR (1992) A comparison between magnetic resonance imaging and computed tomography for stereotactic coordinate determination. Neurosurgery 30:402–407. https://doi.org/10. 1227/00006123-199203000-00015

Chapter 7

Robot-Assisted Deep Brain Stimulation

Catherine Moran and Steven S. Gill

Abstract

In this chapter we explore the application of surgical robotics to Deep Brain Stimulation (DBS). Since the first neurosurgical procedure using robotic assistance was performed, a growing number of neurosurgical indications have emerged and different robotic systems have been developed. In the field of intracranial electrode implantation for deep brain stimulation the ability to first pre-plan trajectories using intuitive planning software, examining and correcting that plan prior to surgery and then executing that plan by maintaining precise targeting and fixed tooling positions during surgery makes robotic techniques an obvious fit for DBS. We look at the advances in neurosurgical robots over the last two decades and describe in detail our experience using the NeuroMate robotic system to implant DBS electrodes.

Key words Deep brain stimulation, Robot, Robotics, Electrode

1 Introduction

The field of Deep Brain Stimulation (DBS) is well suited to the use of robotic surgical techniques. The fundamental objective in implanting DBS electrodes is both high precision and high accuracy of targeting. A robot moves through fixed predetermined motions using kinematic positioning software. These reproducible and accurate movements can then become part of a surgical workflow. Advances in robotic systems together with intra-operative imaging techniques has seen their increasing integration into our operating theatres and their expanding clinical use in DBS surgery.

The first robot used to define the trajectory of a brain biopsy was the Puma (programmable universal machine for assembly) 560 industrial robot (Advance Research and Robotics, Oxford, CT) [1]. As industrial robots were subsequently deemed unsuitable for surgery the development of specific surgical robots began.

In the late eighties and early nineties a group led by Benabid in Grenoble began to develop the first specific neurosurgical robot, which is now the NeuroMate robot. The Neuromate is a robotized arm linked to a stereotactic frame. In parallel, in 1993, the Minerva

Hani J. Marcus and Christopher J. Payne (eds.), *Neurosurgical Robotics*, Neuromethods, vol. 162,
https://doi.org/10.1007/978-1-0716-0993-4_7, © Springer Science+Business Media, LLC, part of Springer Nature 2021

Robot from Lausanne University was used to aspirate cystic intra-cranial lesions followed by brain biopsies in eight patients. This robot was designed to function in the CT scanner and all stages of the surgery were performed by the robot, fully automatically, with the surgeon supervising and checking instrument location or patient status by performing additional CT imaging if necessary during the procedure. Specific tooling for use in it was proposed for development, including probes used for radiofrequency ablation, specifically for use in thalamotomy for the treatment of Parkinson's disease. They designed a three dimensional (side –outlet) electrode probe to allow for a choice of radial directions within the given 'cylinder' or trajectory path. The Minerva project was discontinued prior to widespread use.

Robotic systems used to implant DBS electrodes today include ROSA (Medtech), Pathfinder (Prosurgics/Armstrong Healthcare Ltd. Coleraine, UK) and NeuroMate. In this chapter we will describe the implantation of DBS electrodes using the NeuroMate system, which was the first robotic device for use in neurosurgery to obtain EC brand in Europe and FDA-approval in the US.

2   Summary of Evidence

In 2002 Li et al published in-vitro accuracy data of the NeuroMate robot using a phantom setup [2]. The group quantitatively compared NeuroMate accuracy with three different established stereotactic localization systems (ZD stereotactic ring, LR tracking system registered with bone-screw markers, and CR tracking system registered with the ZD ring). They concluded that the frame-based system was comparable to that of standard frame-based or infrared localising systems. They reported application accuracy of 1.29 mm.

Once preclinical accuracy was determined the first large patient series of implanted DBS electrodes using the NeuroMate was published in 2003 [3]. Varma et al described use of the NeuroMate robot with frameless registration for the treatment of Parkinson's Disease. They evaluated clinical outcomes in 51 patients and documented improvement of 43% in UPDRS III scores at 6 months and 51.7% at 18 months. By 2006, Benabid and his group had used the Neuromate robot for over 500 DBS cases and 2000 SEEG cases [4].

In 2012 LeFranc et al described the use of flat panel CT imaging and robotic implantation of DBS electrodes using the ROSA robot [5]. In 2014 the group subsequently reported accuracies of 27 frame-based (DBS electrode implantations) and

31 frameless stereotactic surgeries and reported targeting accuracy of 0.8 and 1.2 mm respectively using either frame-based or frameless registration [6]. Using a custom phantom device to examine targeting accuracies they also compared varying registration techniques and varying image types used for reference. They found the most accurate targeting was achieved using CT rather than MRI, regardless of registration type (0.3 mm). They also reported clinical improvement in UUPDRS after 6 months of stim following robotic electrode implantation of 30.8 (pre-op 'off med') to 14.15 (post-op 'on stim/off med'). Similar to the NeuroMate robot, the surgeon used the robot to guide the instruments and then manually passes them.

In 2015 Von Langsdorff et al examined in-vivo accuracies of electrode implantation for DBS using the NeuroMate robot in 30 basal ganglia targets and reported mean application accuracy of 0.86 mm, with a maximal error of 1.55 mm [7]. The first case of robotic assisted electrode implantation in the United States was published in 2017, using the ROSA robot, with bilateral implantation of STN electrodes for a 56 year old patient with Parkinson's disease [8].

3 Description of Robot

The NeuroMate is an example of a of supervisory-control robot, which identifies the correct trajectory in 3D space and then provides passive mechanical guidance for the surgeon to perform the procedure (Fig. 1). The NeuroMate has five joints with 6 degrees of freedom, of which 5 rotational and one is linear. Its standard dimensions are 125 cm $\times$ 70 cm $\times$ 125 cm, (L $\times$ W $\times$ H); the arm measures 9 cm $\times$ 8 cm, and it weighs 180 kg.

The arm of the robot can achieve a given position using different specific arm orientations. Calibration of the robot involves calculating the radius of the sphere of the final position of a pointer tip using each different possible arm or joint location which could make that position and calculating their positions with the actual target 3D position. Nineteen different arm positions are used for this calibration, each final location is included in a sphere, and the centre of this sphere is then accorded as the target.

In the DBS procedure the robot first assists in forming the burr hole in the desired and planned location and then assists the introduction of dilators and finally the guide-tube and subsequent electrode to that location.

Both frameless and frame based systems are available, but we describe the use of a frameless system which consists of an ultrasound system used to localise the patients head in space.

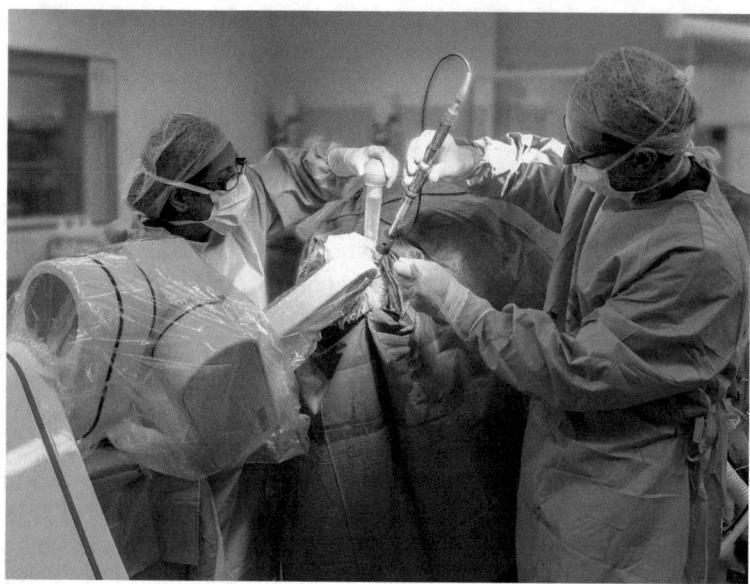

Fig. 1 Use of Neuromate robot during drilling. © Renishaw plc

4 Description of Technique

4.1 Pre-operative Planning

For surgical planning patients undergo 3T MRI (high resolution coronal T2 sequence: 3D TSE, flip angles 90°/100°, TR/TE 1550 ms/100 ms, resolution 0.6 × 0.6 × 0.6 mm; a high resolution magnetization prepared axial T1 sequence 3D TFE, flip angle 8°, TE/TR/TI 5.4 ms/12 ms/458.8 ms, resolution 0.7 × 0.7 × 0.7 mm), and a post gadolinium magnetization prepared axial T1 sequence (3D TFE, flip angle 13°, TE/TR/TI 3.8 ms/8.3 ms/973 ms, resolution 1.0 × 1.0 × 1.0 mm). A volumetric CT angiogram with bony volumes (64 slice, 0.625 mm slice thickness, 120 kV/mA, mim:80, max:450) is also performed. These DICOM image sets are uploaded to NeuroInspire software system and fused. The fusion can be performed using both region of interest (ROI), incorporating the blood vessels at the circle of Willis, and full volumes.

The topographic images of the STN are used to plan the electrode target as well as the approach. A 3D simulation of an electrode is employed in the planning software. Once this 'virtual' electrode is placed at the desired location in the STN, the base sequence was changed to T1 post contrast images allowing for entry and trajectory planning through the brain parenchyma, avoiding vasculature. Fine adjustments using co-ordinate or arc and ring changes can then be performed for final virtual placement of the electrode. Once completed the sequence is changed to the bony volume CT and the software provides wire-frame images of each cranial drill which are each planned in turn.

4.2 Electrode Implantation

Under general anaesthesia the Leksell frame (Elekta Instruments AB, Stockholm) is applied to the patient. The Leksell frame is then used as a head holder to fix the patient to the stereotactic pedestal. The approach is generally pre-coronal with the patient positioned in the head holder with an element of neck extension to allow for minimal obliqueness at entry point.

Once the patient is fixed to the Robot, registration is performed using the Neuro|LocateTM (Renishaw plc, Wotton-under-Edge, UK) system, a five-fiducial marker reference system that is kinematically attached to the robotic arm. A high-resolution 3D full volume CT (gantry rotation $= 360°$, digital flat panel detector $= 40 \times 30$ cm, camera resolution $= 2000 \times 1500$ (3 mega-pixels), pixel pitch $= 0.194$ mm, reconstruction matrix $= 512 \times 512 \times 192$, exposure factors: 120 kv 149mAs) is then taken and used to co-register the robotic stereotactic space to the planning scans, specifically the pre-operative CT angiogram. Intra-venous contrast is used to highlight intracranial vessels to improve registration accuracy. Marker registration is performed automatically, with a manual verification of accuracy post registration.

A kinematic laser tool-pointer is then attached to the robotic arm and moved to mark bilateral entry points on the scalp. The patient and robotic arm are prepped and draped. Bilateral skin incisions are made just behind planned entry points to avoid the guide tube hubs lying directly under the healing wound postoperatively. The approach is generally pre-coronal, just posterior to the hairline, which allows for minimal obliqueness at entry point.

Using the software in delivery mode the laser pointer is replaced with a kinematic instrument guide and moved to the datum position, along the predefined implant position. A skull verification instrument is then passed through the instrument guide and its distance from the skull is input into the software to allow for any fine adjustments of the robotic arm needed prior to drilling.

First, a small pilot-drill (1.85 mm diameter) with a pre-set stop feature is used to breach the inner Table. A gentle 'pecking' motion perpendicular to the skull surface is used to begin the pilot burr hole to minimise any element of slip. Following this, a second core drill (4.6 mm diameter) is passed through the robot tool holder and created a precision location feature for the subsequent dilation and delivery instruments to fit into, as well as a feature for anchoring the press-fit guide-tube hub. At the core drill tip is a non-cutting guidance feature which locates into the previously drilled pilot hole and functions to keep the feature concentric along the planned trajectory. The final precision drill is then used in the same manner to create a datum feature on the skull surface. The datum drill (5 mm diameter) also possesses a guidance, non-cutting, feature which sits in the core drilled hole, again ensuring concentric drilling. This final drill created a 1 mm deep recess. The location of the datum is then verified by passing a circular measuring instrument through the robot guide, similar to the first step described. This value is input into the software and calculation of the dilation and delivery instrument lengths and guide tube lengths are calculated. During the drilling phase liberal saline wash is used at the burr-hole site to prevent CSF egress. A small, 1.8 mm diameter rigid probe with a 5 mm long taper and 0.5 mm tip-radius is then inserted through the small cranial guidance hole, and functions to pierce the dura and superficial cortex.

After adjusting the tooling lengths as per the planning software, a second, much longer probe or fine, blunt dilator, is delivered to the intended target of the DBS electrode. A circular feature at the end of this probe allows it to locate exactly in the precision hole created by the guided core-drill. Using a thumb screw the probe is locked into position in the instrument guide, effectively coupling the patients head to the robot arm. The probe is 1.3 mm diameter, with a 20 mm taper that extends from a 0.5 mm tip radius and is inserted down the track with swift axial rotations. A second probe is then used after being locked in position in the skull guidance feature, with a 1.7 mm diameter and bullet tapered profile, extending 5 mm from a 0.5 mm tip. This probe is used to dilate the previous track to the same diameter as the implantable guide-tube.

Using a variable length cutting jig, the planned guide-tube length (outputted from the software plan) is set and cut, the stylet is cut with it to a length 1 cm longer. The loaded guide-tube delivery tool is then passed through the robot instrument guide and the implantable guide-tube accurately directed through the skull guidance features. The final positioning of the implantable guide tube is confirmed by press-fitting the guide-tube hub on to the machined skull datum feature, where the external treaded feature of its hub anchors it in its corresponding precision drilled hole. Once the delivery tool is removed, a radiopaque stylet of the same diameter as the DBS electrode is delivered through the implantable guide-tube to the DBS target position. The skin incision is then closed temporarily with nylon suture and the robot arm moved to the predefined second entry point. The process is then repeated for the contralateral guide tube and stylet.

Confirmation of stylet positioning as compared to the planned electrode position is achieved by taking a second high-resolution 3D full volume CT using O-arm (Medtronic Inc., Minneapolis, Minnesota, USA) and co-registering with pre-implant imaging. This is obtained after implantation of both guide tubes and stylets, with the patients maintained in the same fixed position on the robotic pedestal.

Once targeting accuracy is confirmed the Leksell head frame is detached from the robot and removed from the patient's head. The patient is re-draped and prepped for both IPG and electrode placement. The depth of the DBS lead, defined by the software, is set by creating a stop from a short length of plastic lead fixation tubing that comes with the lead sets. The lengths can be adjusted according to the stylet location if there is any depth error. The stylets are removed from the guide tubes and the leads are implanted to their stops which engage with the dome shaped hubs of the guide tubes. Whilst keeping the stops engaged with the domed hubs the leads are angled through 90° and fixed to the skull with mini plates. The leads are then connected to extension leads which were tunnelled to a sub-clavicular pocket and finally connected to the IPG.

5 Tips and Tricks

The three factors which contribute to the overall accuracy of robotic intracranial targeting are: first, the mechanical accuracy of the robot; second, the imaging used and the method of their acquisition; and third, the registration technique.

In DBS surgery, perhaps the most significant source of inaccuracy is brain shift that occurs during electrode implantation. Every attempt must be made to mitigate this by minimising CSF egress and pneumocephalus during the procedure by using graduated drills, small burr holes, and graduated, thin parenchymal dilators.

6 Conclusions

Robot-assisted DBS allows for both high precision and high accuracy. However, the Neuromate robot remains large, bulky, and costly. In the future we anticipate smaller, and less costly robots will be developed, allowing for their more widespread use.

References

1. Kwoh YS, Hou J, Jonckheere EA, Hayati S (1988) A robot with improved absolute positioning accuracy for CT guided stereotactic brain surgery. IEEE Trans Biomed Eng 35 (2):153–160. https://doi.org/10.1109/10.1354

2. Li QH, Zamorano L, Pandya A, Perez R, Gong J, Diaz F (2002) The application accuracy of the NeuroMate robot—a quantitative comparison with frameless and frame-based surgical localization systems. Comput Aided Surg 7 (2):90–98. https://doi.org/10.1002/igs.10035

3. Varma TR, Eldridge PR, Forster A, Fox S, Fletcher N, Steiger M, Littlechild P, Byrne P, Sinnott A, Tyler K, Flintham S (2003) Use of the NeuroMate stereotactic robot in a frameless mode for movement disorder surgery. Stereotact Funct Neurosurg 80(1-4):132–135. https://doi.org/10.1159/000075173

4. Benabid AL, Deuschl G, Lang AE, Lyons KE, Rezai AR (2006) Deep brain stimulation for Parkinson's disease. Mov Disord 21(Suppl 14): S168–S170. https://doi.org/10.1002/mds.20954

5. Lefranc M, Le Gars D (2012) Robotic implantation of deep brain stimulation leads, assisted by intra-operative, flat-panel CT. Acta Neurochir 154(11):2069–2074. https://doi.org/10.1007/s00701-012-1445-7

6. Lefranc M, Capel C, Pruvot AS, Fichten A, Desenclos C, Toussaint P, Le Gars D, Peltier J (2014) The impact of the reference imaging modality, registration method and intraoperative flat-panel computed tomography on the accuracy of the ROSA(R) stereotactic robot. Stereotact Funct Neurosurg 92(4):242–250. https://doi.org/10.1159/000362936

7. von Langsdorff D, Paquis P, Fontaine D (2015) In vivo measurement of the frame-based application accuracy of the Neuromate neurosurgical robot. J Neurosurg 122(1):191–194. https://doi.org/10.3171/2014.9.JNS14256

8. Vadera S, Chan A, Lo T, Gill A, Morenkova A, Phielipp NM, Hermanowicz N, Hsu FP (2017) Frameless stereotactic robot-assisted subthalamic nucleus deep brain stimulation: case report. World Neurosurg 97:762.e711–762.e714. https://doi.org/10.1016/j.wneu.2015.11.009

Chapter 8

Robot-Assisted Stereoelectroencephalography Implantation

Aswin Chari, M. Zubair Tahir, and Martin M. Tisdall

Abstract

Stereoelectroencephalography (SEEG) plays a key role in the pre-surgical evaluation of patients with drug resistant epilepsy. It involves the insertion of multiple (10–20+) electrodes into the brain parenchyma to attempt to identify the epileptogenic zone that may subsequently be resected to achieve seizure freedom. Increased safety and accuracy afforded by modern technology has made SEEG more acceptable as a diagnostic procedure in the evaluation of both adults and children with focal epilepsy. In this chapter, we outline the indications, history and evidence for SEEG and describe the robot-assisted technique used at our institution.

Key words Stereoelectroencephalography, SEEG, Epilepsy, Robotics

1 Introduction

The main aim in the evaluation of patients for epilepsy surgery is identification of the epileptogenic zone (EZ), which can be defined as *'the minimum amount of cortex that must be resected (inactivated or completely disconnected) to produce seizure freedom'* [1]. The majority of patients will have this epileptogenic zone identified by non-invasive methods including history and clinical examination, imaging and electrophysiological evaluation (Table 1).

In a subset of patients, there will be sufficient uncertainty to warrant further, more invasive evaluation with intracranial EEG recordings. There are four main indications that prompt the epilepsy multidisciplinary team (MDT) to advocate for intracranial recordings [2]:

1. Lesion-negative cases: There is no clear lesion identified on high-resolution MRI. Although the non-invasive work-up may identify the hemisphere or even lobe, sub-lobar localization from scalp EEG is often poor and invasive monitoring is required.

Hani J. Marcus and Christopher J. Payne (eds.), *Neurosurgical Robotics*, Neuromethods, vol. 162,
https://doi.org/10.1007/978-1-0716-0993-4_8, © Springer Science+Business Media, LLC, part of Springer Nature 2021

Table 1
Key facets of the non-invasive pre-operative evaluation of patients for epilepsy surgery

History and clinical examination	Seizure semiology and frequency Comorbidities Clinical neurological examination Neuropsychologic and neuropsychiatric evaluation
Imaging	High resolution magnetic resonance imaging (MRI) Positron emission tomography (PET) scan Ictal single-photon emission computed tomography (SPECT) scan Functional MRI (fMRI)
Electrophysiological evaluation	Scalp electroencephalography (EEG) Video-EEG monitoring Magnetoencephalography (MEG)
Combined approaches	EEG-fMRI ESI (electrical source imaging)

2. Lesion-positive cases in which there is discordance between the imaging and non-invasive electrophysiological localization.

3. Lesion-positive cases which are adjacent to eloquent cortex: In some instances, invasive recording may be needed to identify the boundaries of the epileptogenic zone and eloquent cortex to determine the risk benefit balance of resection of the EZ.

4. Cases with multiple lesions in which the electrophysiological investigations are not able to adequately ascertain which lesion (s) are responsible for seizure generation.

Extraoperative intracranial recordings come in two main forms, namely subdural grid or strip recordings and stereoelectroencephalography (SEEG) depth electrodes. Choice of technique will vary between centres however in our institution grids and strips are generally used to evaluate patients with clear cortical lesions, where the main aim of the invasive recording is to identify boundaries between the lesion and eloquent cortex. Grids are less useful when there are deep or difficult to access (e.g. interhemispheric surface, insula) lesions, multiple lesions or lesion negative cases. Therefore, the mainstay of invasive evaluation is by SEEG recordings, which allows a 3-dimensional assessment of epileptogenic networks. However, SEEG also has its limitations, mainly the reduced density of cortical sampling leading to poorer functional localisation; lesions next to eloquent cortex that are identified through SEEG therefore may require additional adjuncts for functional localization such as pre-operative fMRI and diffusion tensor imaging (DTI), additional subdural grid or strip mapping or intraoperative mapping using awake surgery with cortical stimulation.

The choice of whether to proceed with SEEG monitoring is complex and requires an epilepsy MDT team and the involvement of the patient and their family, who must have the ability to engage in complex discussions about the likelihood of the SEEG process to identify a resectable EZ and the subsequent likelihood of a resection to provide acceptable outcomes in terms of reduction/elimination of seizures and preservation of neurological function. Patients (or relatives) who are unable to partake in such discussions or patients who may be unable to comply with having intracranial electrodes may not be candidates for invasive recordings.

1.1 Historical Perspective

During the early twentieth century, advances in understanding the importance of EEG recordings (both invasive and non-invasive) in the localization of both normal function and abnormal epileptogenic tissue led to the birth of modern epilepsy surgery, pioneered by Penfield and Jasper in Montreal [3, 4]. The concept of SEEG recording for focal epilepsy was first reported in 1949 [5] but was limited by the lack of a stereotactic atlas. Individualised SEEG using a stereotactic map guided by pneumencephalogram was developed in France in the 1950s by Jean Talairach and Jean Bancaud [6, 7]. The procedure was well tolerated, which allowed longer-term recordings over days and therefore ictal recordings that allowed localization of an epileptogenic focus [8]. The term SEEG was coined in 1962 [6].

The use of invasive recordings declined in the 1980s following the advent of CT, MRI, functional imaging such as PET and ictal SPECT and prolonged telemetric video-EEG monitoring [9]. The use of invasive recording therefore became more limited, to encompass the 4 indications outlined above. Current practice has been streamlined by improved technology including frame-based and frameless neuronavigation techniques, robot-assisted surgery and the more nuanced use of combined grid and depth electrode recordings [10].

A comprehensive review of the history of SEEG is provided in this review [9].

2 Summary of Evidence

Several variations exist for both manual and robot-assisted implantation procedures, using both frame-based and frameless approaches. Each system carries its own set of advantages and disadvantages and, irrespective of the system used, the ultimate aim of the operation is to achieve safe and accurate implantation.

The main potential adverse events are haemorrhage, infection, hardware-related complications and neurologic sequelae. A meta-analysis of SEEG related complications over the last 30 years revealed a pooled prevalence of complications as listed in Table 2

Table 2
Pooled prevalence of complications from SEEG from recent meta-analysis [11]

Complication	Pooled prevalence (%)	95% confidence interval (%)
Haemorrhage: Including intracerebral, subdural and extradural haematomata	1.0	0.6–1.4
Infection: Including intracranial abscess, meningitis and superficial wound infections	0.8	0.3–1.2
Hardware failure: Including fracture, recording malfunction and accidental removal	0.4	0.0–0.7
Neurologic Sequelae: Including permanent motor or sensory deficits	0.6	0.2–1.0
Death	0.3	0.1–0.6
Overall complications	1.3	0.9–1.7

[11]. In recent years, advancements in planning software have allowed safer planning of trajectories, incorporating CT, MR or conventional angiographic data to avoid blood vessels, thereby reducing the risk of haemorrhage, neurological complications and death.

Accuracy can be assessed in several ways including deviation from the planned entry point, distance from the planned target and angular deviation from the planned trajectory; there is as yet no standardised way of assessing this accuracy. A recent systematic review found 15 studies that allowed assessment of the comparative accuracies of these different approaches [12]. Due to the heterogeneity of study designs and endpoints used to assess accuracy, no meaningful comparisons could be made. However, the available low-level evidence did point towards increased accuracy and speed of robot-assisted techniques. It was not possible to assess comparative complication rates. The authors of this study are currently conducting a randomised trial to compare the accuracy, speed and complication rates of robot-assisted and manual SEEG implantation procedures using a frameless system [13].

3   Description of Robot

At our institution, all procedures are conducted using a frame-based approach (Leksell® frame) with robotic assistance from the Renishaw neuromate® stereotactic robot (Fig. 1). This robot, currently in its third generation is specifically designed for neurosurgery and has been used in over 10,000 procedures. It has FDA approval (USA) and is CE marked (Europe) for clinical use.

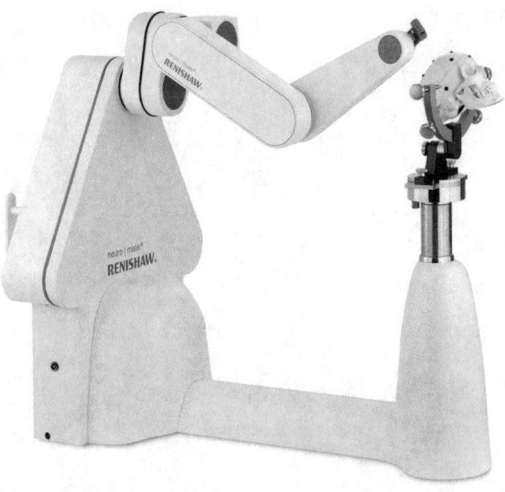

Fig. 1 Renishaw neuromate® stereotactic robot

The neuromate® stereotactic robot is a compact system that fits easily into an operating theatre. It functions via a supervisory-controlled system in which the surgeon plans a safe surgical trajectory using pre-operative volumetric imaging, it registers using frame-based or frameless methods and the robot then carries out trajectory alignment autonomously under the supervision of the surgeon. It has 6 joints that affords it 6 degrees of freedom, with the possibility of arranging the joints in multiple configurations to facilitate access to multiple trajectories. It has been widely used for SEEG, deep brain stimulation, stereotactic biopsy, neuroendoscopy and experimental procedures such as implantation of convection enhanced drug delivery catheters [14].

4 Description of Technique

4.1 Planning of Electrode Implantation

Planning of electrode locations begins at the epilepsy MDT meeting, with a thorough discussion of the case including review of the patient's medical history, seizure semiology, psychological and psychiatric evaluations, imaging and non-invasive electrophysiological evaluations. This generates a hypothesis as to where the seizures may be coming from and a rationale for implantation of SEEG electrodes. Electrode trajectories are suggested by the treating clinical neurophysiologist and debated with the entire team, with the aim of definitively establish the source of the seizures whist minimizing the number of electrodes to maximize safety. The procedure may be planned as a unilateral or bilateral implantation depending on the information from the pre-operative evaluation.

Planning is conducted on the neuroinspire™ surgical planning software that integrates with Renishaw neuromate® stereotactic

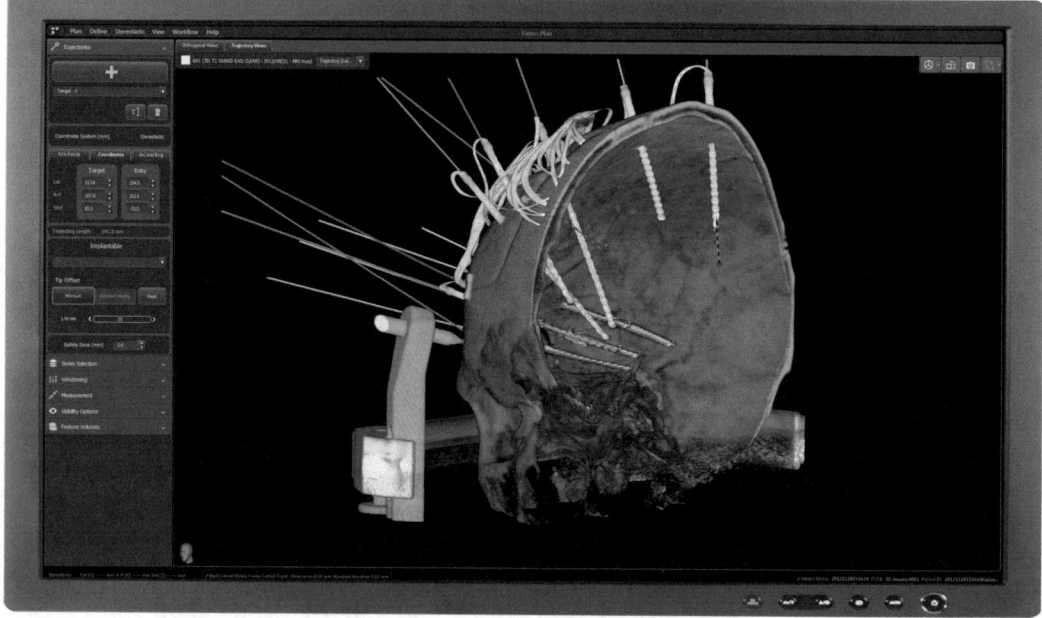

Fig. 2 Example of the intraoperative workstation with neuroinspire™ surgical planning software that integrates with the Renishaw neuromate® stereotactic robot

robot (Fig. 2). Multiple MRI (T1, T1 post contrast, T2 and Flair) and CT (high resolution CT angiography with commercial bone stripping algorithm from Siemens) sequences are used to calculate safe electrode trajectories that avoid surface and sulcal vessels and maximize coverage of cortical grey matter. The software is able to render 3D volumes to aid visualization and ensure that the trajectories do not cross or clash.

4.2 Surgical Procedure

In the anaesthetic room, under general anaesthesia, a Leksell® frame is fitted, ensuring that the frame is placed sufficiently low to avoid compromise of electrode trajectories. A skin staple is placed just behind the hairline as a fiducial marker. The patient is taken to the CT-scanner for a non-contrast volumetric CT scan that is co-registered to the pre-planned trajectories on the neuroinspire™ software. This registers the coordinates of the electrode trajectories with respect to the frame.

On return to the operating theatre, the patient is placed supine on the operating table and the Leksell® frame is fixed to the neuromate® robot (Fig. 3a). Care must be taken not to move the operating table once the frame is fixed to the robot as the patient's head will not move with the table; as a safety measure, once in position, the operating table remote control is disconnected. The registration is then checked for accuracy by verifying a laser along a robotic trajectory to the previously affixed staple. If necessary, this can be used to re-check registration accuracy during the procedure.

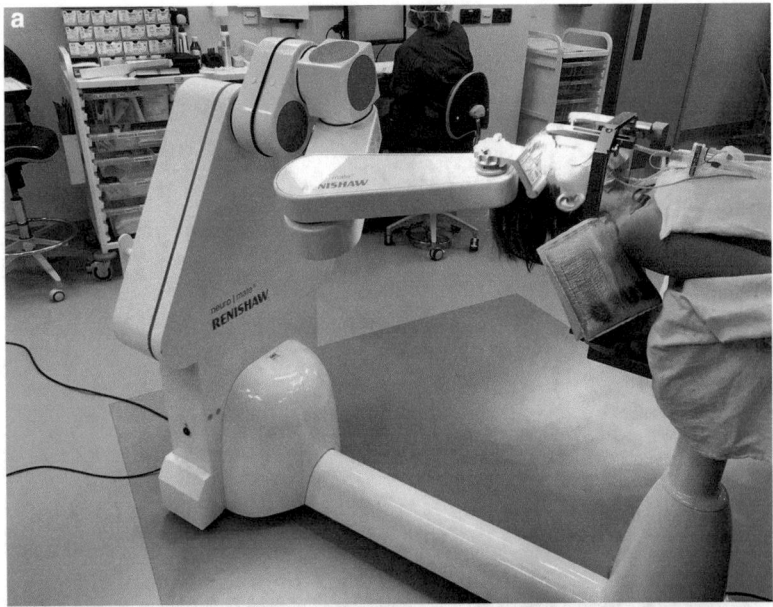

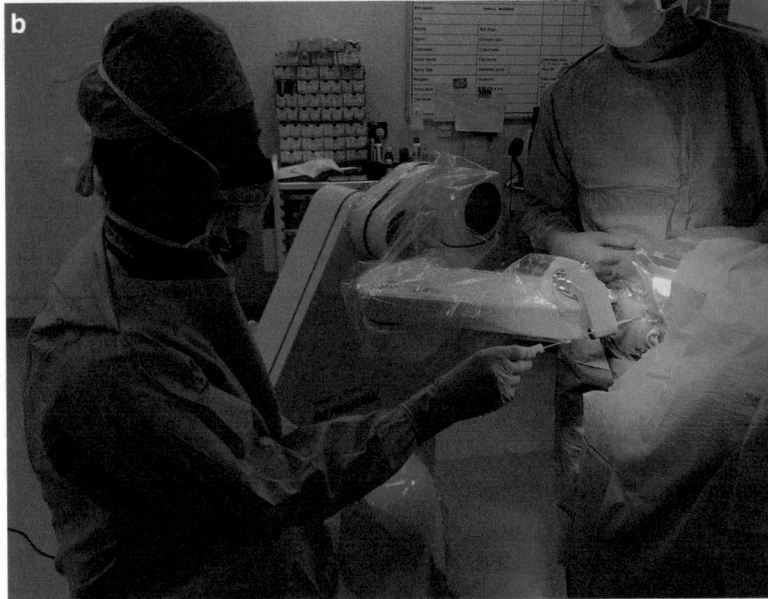

Fig. 3 Surgical workflow. Patient is positioned on the operating table with the frame attached directly to the robot (**a**). Following prepping and draping, for each electrode, a stab incision is made using bone spike (**b**), a burr hole is made using electric drill (**c**) and durotomy is made using monopolar cautery (**d**). The bolt is screwed into the skull (**e**) and the electrodes are placed once all the bolts are secured (**f**). The final result following implantation of all 11 electrodes is shown (**g**)

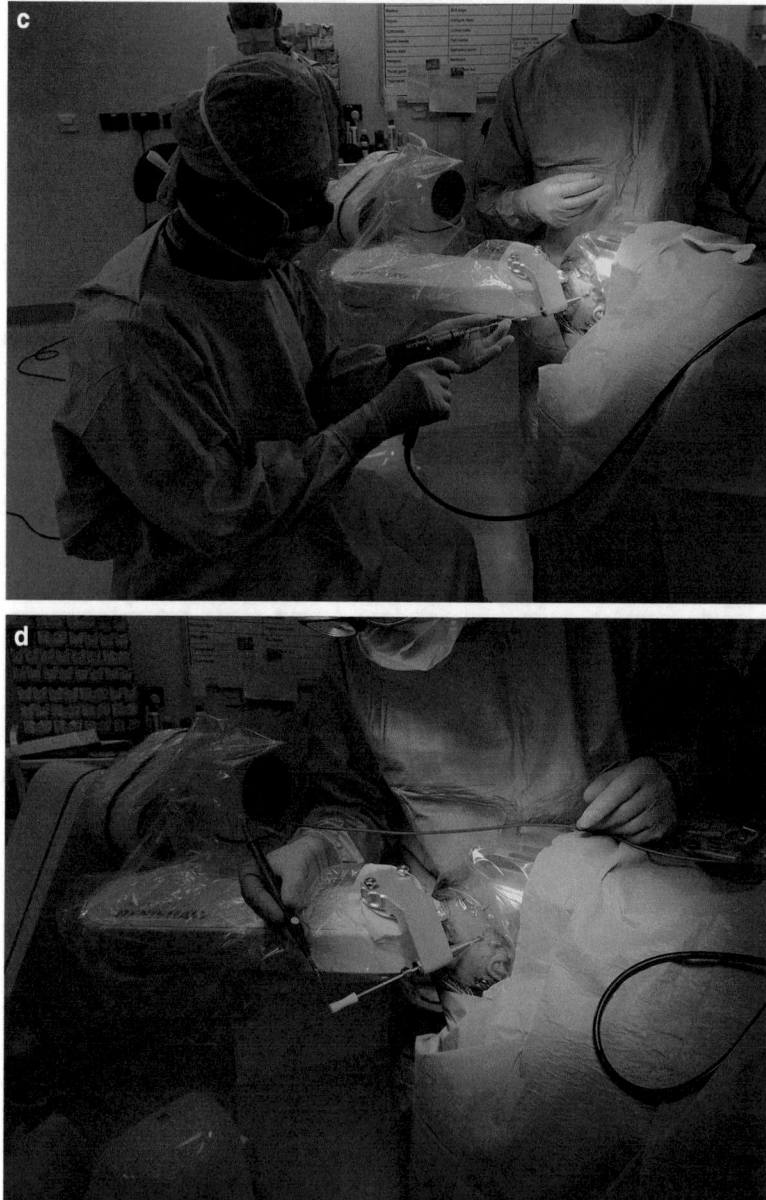

Fig. 3 (continued)

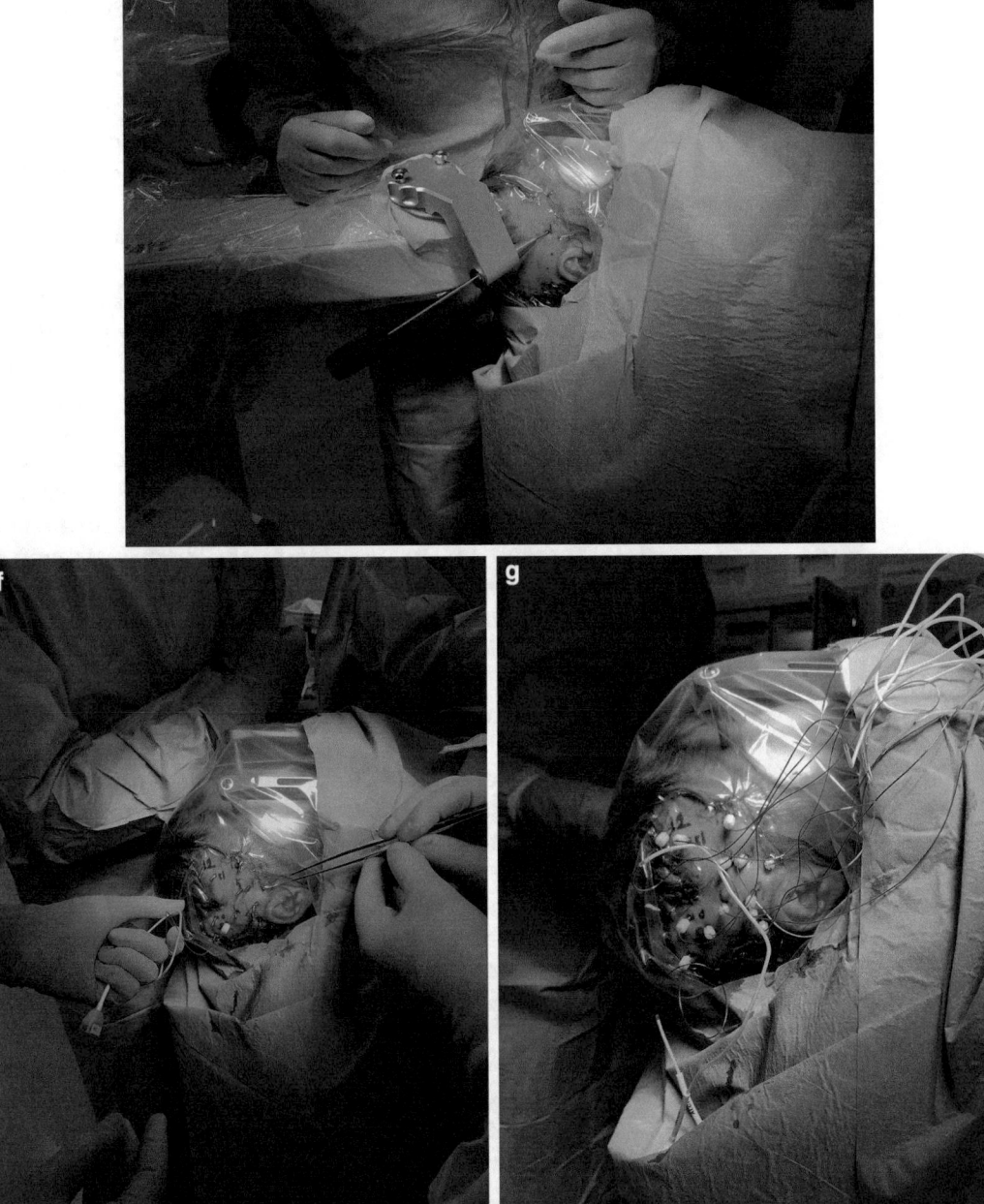

Fig. 3 (continued)

A sphere of safety is created to inform the robot of the location of the patient's head and avoid the arm clashing with this space. The robotic arm is then moved to each of the trajectories to ensure that the arm can access each trajectory; the entry points are marked. The patient is then shaved to expose the entry points, prepped and draped (Fig. 3b). The robotic arm is also draped and a sterile guide tool is attached that can accommodate the required instruments.

Each trajectory is then accessed in turn and a skin incision, twist drill hole and durotomy with monopolar are performed before the bolt is screwed into the skull (Figs. 3c–f). The drill's safety stop position is calculated pre-operatively for each electrode to reach no further than the inner table of the skull. Care is taken to minimise CSF loss in order to reduce brain shift. A rod is used to measure the distance between the robotic arm and the tip of the anchor bolt, allowing the total length of the electrode to be calculated. Once all the bolts have been inserted, a stylet is inserted to the pre-defined depth and the electrode is then inserted and secured to the bolt (Fig. 3g). Throughout the procedure, the electrode number, corresponding location and active & total lengths are displayed on a spreadsheet in multiple locations around the operating theatre to ensure that each electrode is positioned correctly and inserted to the pre-defined depth. The total number of electrodes varies between cases but is usually between 10 and 20 (Fig. 3h).

4.3 Post-operative Care

Following completion of the procedure and recovery from general anaesthesia, the patient is transferred from the recovery area to the CT scanner to obtain a post-operative non-contrast volumetric CT scan. This allows assessment of post-operative complications and is then registered to the pre-operative planned trajectories to assess accuracy of implantation.

At our institution, perioperative antibiotic prophylaxis is given according to the local trust guidelines but this is not continued routinely during the period of implantation.

Once the patient has been sufficiently monitored (usually 6–10 days, capturing at least 3 seizures of typical semiology), the patient is brought back to the operating theatre and the electrodes and bolts are removed under general anaesthesia with a single suture placed at each exit site. The patient is then discharged following a further 24 h of neurological observation.

Data from the monitoring period is then analysed off-line and discussed at the epilepsy specialist MDT meeting where a decision is made about potential resectability of the epileptogenic zone. These findings are subsequently discussed with the parents and child in the outpatient clinic and, if appropriate, a surgical resection is then offered.

5 Tips and Tricks

1. Planning is key: Irrespective of how accurately the trajectories are adhered to during the implantation, the safety rests upon on meticulous planning using angiographic sequences to ensure that blood vessels are avoided. Based on our collected accuracy data we ensure no vessels are seen within a 5 mm diameter safety zone around each electrode.

2. Fitting the frame: Take care when fitting the frame to ensure none of the trajectories will be compromised by the posts. Fit inferiorly enough to allow the low temporal electrodes (usually to amygdala and hippocampus) to be inserted. Extra attention must be paid to the posterior posts when occipital electrodes are required.

3. The staple: Having the staple to confirm registration accuracy is an important safety check both at the beginning and, if required, during the procedure.

4. Laser marking: Check that the robot can achieve all the electrode trajectories before starting by using the laser attachment. If necessary, the arm configuration of the robot can be altered to achieve the trajectories. Doing this prior to prepping and draping avoids the complexity of having to adjust during the procedure when both the patient and the robotic arm are draped.

5. Check the depth: Each electrode needs to be inserted to the correct pre-planned depth. Check and check again prior to inserting!

6 Conclusions

SEEG is a crucial tool in the armamentarium of the epilepsy MDT in the pre-surgical evaluation of patients with medically refractory focal epilepsy. Advances in robotic-assisted techniques and computer-assisted planning have made the procedure safer and more accurate. The indications for SEEG may broaden, given the potential to combine this diagnostic procedure with lesioning techniques such as radiofrequency thermocoagulation and laser interstitial thermal therapy [15, 16].

We believe our technique to be safe and accurate, whilst acknowledging that there are many alternative systems, both for computer-assisted planning and robotic-assisted surgery, that offer similar safety and accuracy. Ultimately, the safety of planning and execution of SEEG lies in creating a reproducible workflow that the whole team (surgeons, scrub staff, anaesthetists, neurophysiologists) are familiar with, thus minimising error and risk.

References

1. Luders HO, Najm I, Nair D, Widdess-Walsh P, Bingman W (2006) The epileptogenic zone: general principles. Epileptic Disord 8(Suppl 2):S1–S9

2. Jayakar P, Gotman J, Harvey AS, Palmini A, Tassi L, Schomer D et al (2016) Diagnostic utility of invasive EEG for epilepsy surgery: indications, modalities, and techniques. Epilepsia 57(11):1735–1747

3. Penfield W, Steelman H (1947) The treatment of focal epilepsy by cortical excision. Ann Surg 126(5):740–762

4. Jasper HH (1949) Electrical signs of epileptic discharge. Electroencephalogr Clin Neurophysiol 1(1):11–18

5. Hayne RA, Belinson L, Gibbs FA (1949) Electrical activity of subcortical areas in epilepsy. Electroencephalogr Clin Neurophysiol 1(4):437–445

6. Talairach J, Bancaud J, Bonis A, Szikla G, Tournoux P (1962) Functional stereotaxic exploration of epilepsy. Confin Neurol 22:328–331

7. Bancaud J, Dell MB (1959) Technics and method of stereotaxic functional exploration of the brain structures in man (cortex, subcortex, central gray nuclei). Rev Neurol 101:213–227

8. Talairach J, Bancaud J (1966) Lesion, "irritative" zone and epileptogenic focus. Confin Neurol 27(1):91–94

9. Reif PS, Strzelczyk A, Rosenow F (2016) The history of invasive EEG evaluation in epilepsy patients. Seizure 41:191–195

10. Munyon CN, Koubeissi MZ, Syed TU, Luders HO, Miller JP (2013) Accuracy of frame-based stereotactic depth electrode implantation during craniotomy for subdural grid placement. Stereotact Funct Neurosurg 91(6):399–403

11. Mullin JP, Shriver M, Alomar S, Najm I, Bulacio J, Chauvel P et al (2016) Is SEEG safe? A systematic review and meta-analysis of stereo-electroencephalography-related complications. Epilepsia 57(3):386–401

12. Vakharia VN, Sparks R, O'Keeffe AG, Rodionov R, Miserocchi A, McEvoy A et al (2017) Accuracy of intracranial electrode placement for stereoencephalography: a systematic review and meta-analysis. Epilepsia 58(6):921–932

13. Vakharia VN, Duncan J (2016) A randomised control trial of SEEG electrode placement methods. http://www.isrctn.com/ISRCTN17209025.

14. Barua NU, Hopkins K, Woolley M, O'Sullivan S, Harrison R, Edwards RJ et al (2016) A novel implantable catheter system with transcutaneous port for intermittent convection-enhanced delivery of carboplatin for recurrent glioblastoma. Drug Deliv 23(1):167–173

15. Cossu M, Cardinale F, Casaceli G, Castana L, Consales A, D'Orio P et al (2017) Stereo-EEG-guided radiofrequency thermocoagulations. Epilepsia 58(Suppl 1):66–72

16. Cobourn K, Fayed I, Keating RF, Oluigbo CO (2018) Early outcomes of stereoelectroencephalography followed by MR-guided laser interstitial thermal therapy: a paradigm for minimally invasive epilepsy surgery. Neurosurg Focus 45(3):E8

Chapter 9

Robot-Assisted Endoscopic Third Ventriculostomy

Jillian Plonsker and David D. Gonda

Abstract

Endoscopic third ventriculostomy has an established role in the treatment of obstructive hydrocephalus. In the traditional technique, the entry site and trajectory of the endoscope is planned based on anatomical landmarks. The endoscope is then guided and controlled by the surgeon in a freehand manner. Recent studies of ETVs performed in North America report an incidence of unintended thalamic or hypothalamic bruising and forniceal stretch injuries from the freehand method as high as 20%. Robotic stereotactic techniques have been introduced for endoscopic third ventriculostomy with the goal of reducing traction and minimizing manipulation of neural structures by optimizing trajectories and providing mechanical stabilization to the endoscope. In this chapter, we outline our methods and describe the robot-assisted technique used at our institution.

Key words Hydrocephalus, ETV, Endoscopic Third Ventriculostomy, Robotics

1 Introduction

Hydrocephalus is among the most commonly encountered neurological pathologies by adult and pediatric neurosurgeons alike. It results when there is an imbalance in the brain's production and subsequent reabsorption of spinal fluid causing excessive accumulation within the central nervous system and elevation of intracranial pressure. Insertion of ventricular shunts are effective in diverting the excessive spinal fluid but are accompanied by a host of life long challenges related to CSF over drainage, under drainage, infection or failure of implanted hardware. In cases of obstructive hydrocephalus, endoscopic third ventriculostomy provides an attractive alternative to permanent hardware implantation for CSF diversion.

The first endoscopic third ventriculostomy was performed by Mixter in 1923 using urologic instruments to visualize and perforate the floor of the third ventricle creating an alternate pathway for CSF to exit the ventricular system to basilar cistern and the extracerebral space where the fluid is eventually reabsorbed through

Hani J. Marcus and Christopher J. Payne (eds.), *Neurosurgical Robotics*, Neuromethods, vol. 162,
https://doi.org/10.1007/978-1-0716-0993-4_9, © Springer Science+Business Media, LLC, part of Springer Nature 2021

natural mechanisms [1]. The operation has been since refined to include the use of more sophisticated endoscopes, stereotaxy, and adjuvant cauterization of the choroid plexus [2]. The probability of an ETV successfully treating a patient's hydrocephalus can be assessed by considering the patient's age, etiology of hydrocephalus, and whether the patient has a previously implanted ventricular shunt. These factors are weighed and measured on a scale known as the ETV Success Score (ETVSS) [3]. The ETVSS has been validated across multiple countries and hospital systems [4, 5].

In the traditional technique, the entry site and trajectory of the endoscope is planned based on anatomical landmarks. The endoscope is then guided and controlled by the surgeon in a freehand manner. At our institution we have adopted the ROSA to assist in many of our endoscopic third ventriculostomy procedures. Robotic surgical assistance allows the surgeon to plan a stereotactically exact approach trajectory, provides a platform to hold and stabilize the instruments while maintaining maneuverability for intra-operative corrections as needed. Better described as mechanized stereotaxy than an autonomously operating device, robotic surgical assistance applied to intraventricular endoscopy allows for a minimally invasive and accurate operative strategy to improve the procedure from its current freehand techniques. In addition to optimizing accuracy of trajectory and instrument stabilization, the robotic surgical assistance allows for a smaller incision size requiring only a single stitch closure and avoids the need for cutting any hair.

Patient's chosen to undergo ETV are considered eligible for robotic assistance as long as their skulls are mature enough to allow for safe application of the headframe. Some patients with chronic hydrocephalus and resultant macrocrania are excluded due to the limits of our endoscope length (150 mm) which is further shortened about 3 cm by the bushing it passes through held by the robotic arm. When the trajectory length from outer table of skull to floor of the third ventricle exceeds 12 cm length, the endoscope will not reach. Also the narrowness of the burr hole made with robotic assistance limits endoscopic manipulation to only a few millimetres in any direction. The limited range of motion provides protection of adjacent brain structures but eliminates the ability to perform supplemental cauterization of the choroid plexus or biopsy of nearby intraventricular tumour which requires a greater degree of endoscopic manipulation than is afforded by the minimally invasive robotic technique. The final contraindication to robotic assistance is in cases where need for treatment of the hydrocephalus is emergently time sensitive since the ROSA requires a stereotactic registration process which adds about half an hour to the surgery time.

2   Summary of Evidence

Recent data analysis from the Hydrocephalus Clinical Research Network of endoscopic third ventriculostomies performed in North American hospitals discovered that the most common complication was inadvertent stretch injury to the fornices (16.6%). Combined with the incidence of thalamic (1.8%) and hypothalamic (1.5%) contusions, the number of local cerebral injuries related to freehand techniques during ETV is almost 20% [6]. Other reported complications such as pituitary dysfunction, third cranial nerve palsies, or injuries to vascular structures can occur but happen far less often [7, 8]. Strategies to reduce these injuries call for improved planning of approach trajectory through the foramen of Monro as the endoscope passes the fornices and thalamus en route to the third ventricular floor as well as a stabilizing platform for better control of the endoscope to reduce surgical gestures and manipulation once the surgical instruments are in place [7–9].

The initial experience using the ROSA for assisted endoscopic third ventriculostomy in 9 patients at our institution was published demonstrating proof of principle [10]. There were no complications in this series and no incidents of forniceal stretch injuries or thalamic contusions. Since that publication we have gone on to perform more than 50 robotic assisted endoscopic third ventriculostomies with similar results, however a larger comparative study is still needed to prove its effectiveness and safety in comparison to traditional freehand methods.

3   Description of Robot

At our institution, all robotic procedures are conducted using the *Ro*botic *S*urgical *A*ssistant ROSA (ROSA, ZimmerBiomet). ROSA has been applied in minimally invasive neurosurgical techniques since 2007. It has aided neurosurgeons in a multitude of minimally invasive procedures which include lead placement for deep brain stimulators, brain biopsies [11], depth electrode placement for seizure monitoring [12], laser ablation of epileptogenic foci [13, 14], and spinal pedicle screw fixation [15]. This robotic surgical assistant empowers the neurosurgeon to perform precise, stereotactic gestures with the aid of stereotaxy and surgical instrument stabilization. This robot is currently in its third generation and has FDA approval (USA) and is CE marked (Europe) for clinical use.

The ROSA robot has a mechanical arm with six degrees of freedom in an architecture similar to a human's, allowing for haptic surgical movements and freedom in the choice of trajectory (Fig. 1). It was initially designed for SEEG electrode placements but has since been applied to a wide array of neurosurgical

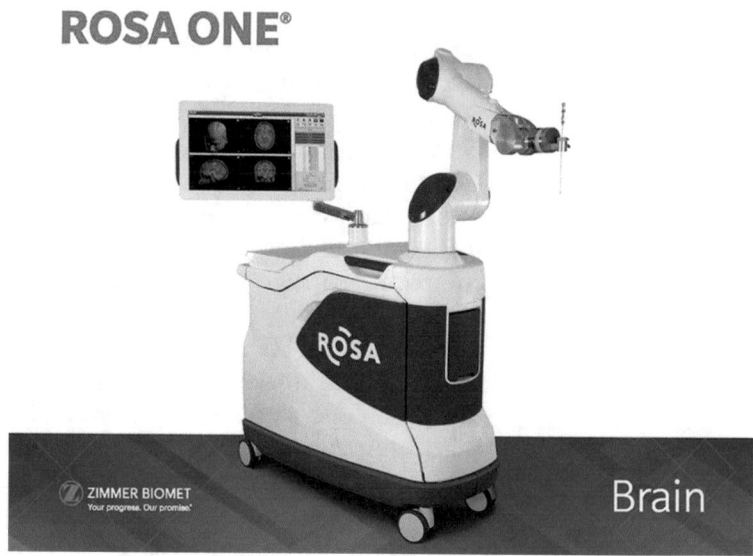

Fig. 1 ROSA® stereotactic robot

procedures including stereotactic biopsies, aspirations, implantation of DBS and RNS electrodes, and insertion of fiberoptic catheters for LITT laser ablations of tumors and epileptic foci. Surgical planning of trajectories is done pre-operatively using thin cut volumetric imaging on the device's software. Registration to the patient intraoperatively is done either with laser scanning of anatomic landmarks (facial features) or by way of touch alignment to surgically implanted in-bone skull fiducials. The robotic arm performs automated movements guiding the surgeon to pre-planned trajectories but is also aided by manual adjustments as is needed during the operation giving the surgeon freedom of movement with the advantages of repeatability and precision.

4   Description of Technique

4.1   Planning of Endoscope Trajectory

The ROSA work station is used for trajectory planning. Thin cut CT or MRI sequences are uploaded into the ROSA system. CT angiograms or T1-weighted MRI of brain with contrast are preferred so that large vessels along the cortex can be avoided however we tend to use whatever imaging is available as long as thin cuts are available for 3D reconstruction. If no stereotactic quality imaging is available, we obtain a new MRI or thin-cut CT sequence.

Trajectory planning is done using the ROSA software (Fig. 2). The trajectory target is labeled on the floor of the third ventricle just anterior to the BA in the sagittal view. A temporary entry point within the foramen of Monro on coronal and sagittal views is then

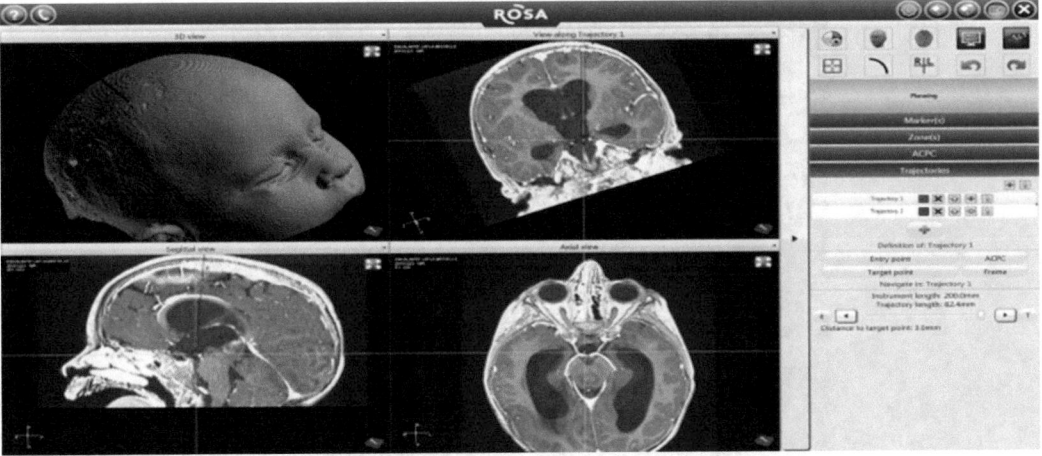

Fig. 2 Example of the intraoperative workstation with ROSA™ surgical planning software

defined. This trajectory is then traced back to the level of the skull, where the entry point is adjusted to a second interim position. The course of the trajectory is then analyzed and minor adjustments are made to avoid any vascular structures along the course. Usually the entry site is very close to the coronal suture 2 or 3 cm away from midline. It is important that the final entry site is positioned at the middle point of the skull thickness, because this will later act as the isometric point during the surgery for any required minor manipulations of the endoscope trajectory. Any additional trajectories are then planned as needed if a secondary procedure was being performed (e.g., needle biopsy).

4.2 Surgical Procedure

In the operating room, under general anaesthesia, a head frame is fitted to hold the skull and securely attached to the ROSA robot (Fig. 3a). We prefer the Sugita or Leksell head frames for robotic stereotactic procedures due to their rigid 4 point-fixation, however a standard 3 pin Mayfield headholder may also be used. The patient lies supine with the head slightly flexed for exposure of the planned incision site. The ROSA's laser attachment is used to register the patients position by scanning the facial landmarks.

In cases in which there are multiple procedures to be performed, we strategically choose to do the one that requires the most accurate stereotaxy first. Therefore, the procedures that require biopsies are performed before the ETV to prevent potential reconfiguration in the ventricular system anatomy that could occur with the ETV and ventricular decompression. Each patient as sterilely prepared and draped in the usual fashion. No head shaving or hair cutting is performed. The ROSA arm guides the surgeon to the decided entry site on the scalp, where a No. 15 blade is used to make a single stab incision through the skin (Fig. 3b), through

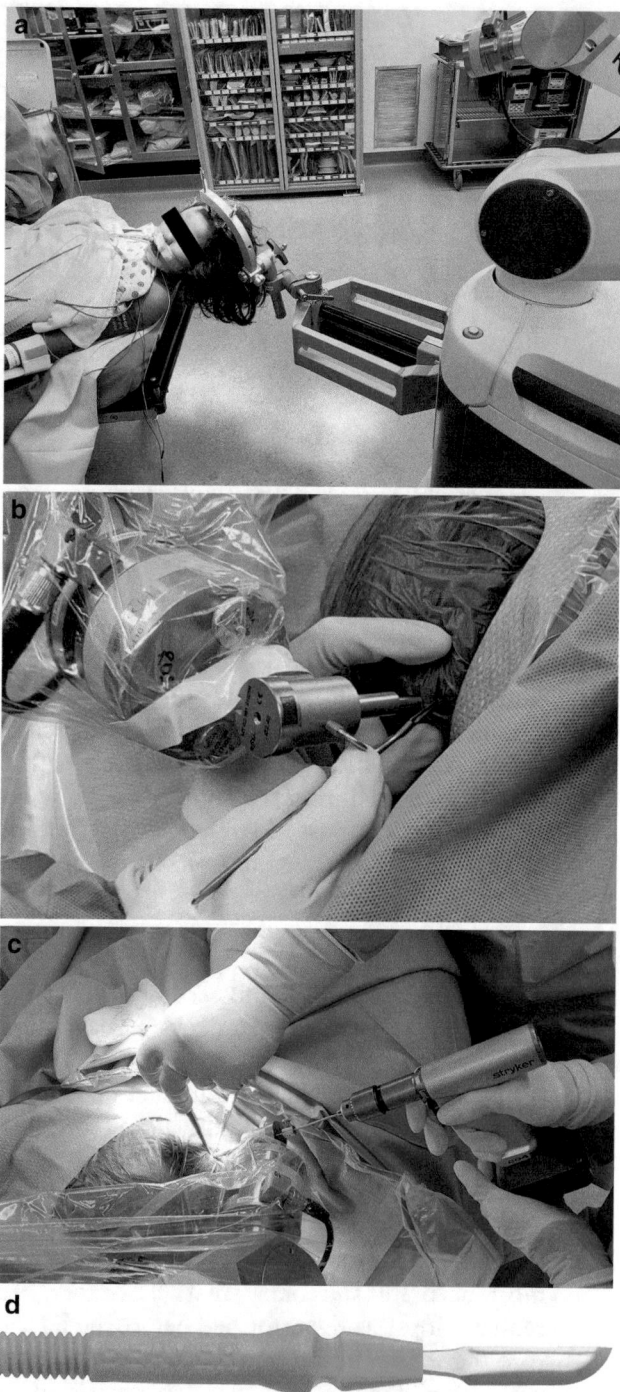

Fig. 3 Surgical workflow. Patient is positioned on the operating table with the frame attached directly to the robot (**a**). Following prepping and draping, a stab incision is made using a 15 blade (**b**), a 3.2 mm diameter burr hole is made using electric drill (**c**) and durotomy is made using #67 Beaver blade (**d**). The Endoscope is passed into the ventricular system (**e**) and the floor of the third ventricle is fenestrated (**f**). The final result showing visualization of the naked basilar artery (**g**)

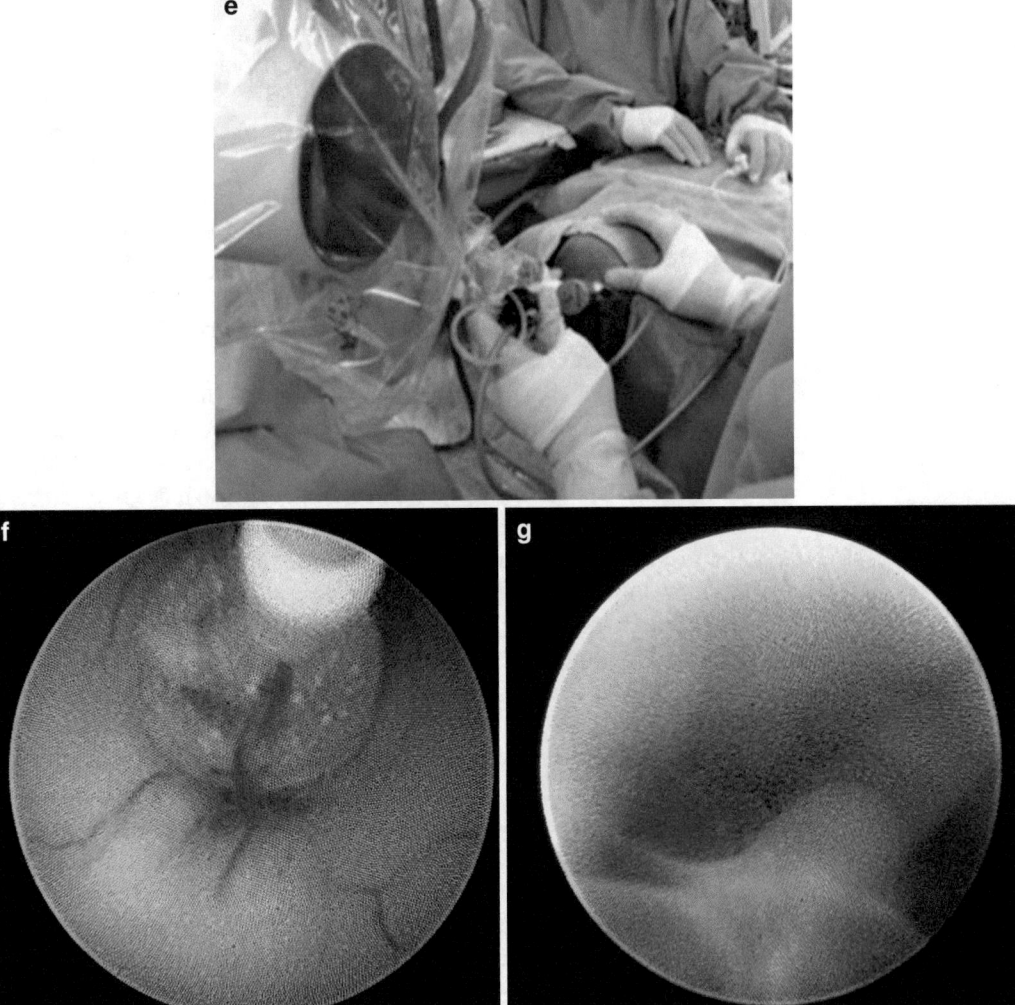

Fig. 3 (continued)

which we passed a 3.2-mm-diameter drill bit to make a bur hole through the skull along the trajectory (Fig. 3c).

A No. 67 Beaver blade small enough to pass through the burr hole is used to make a cruciate incision through the dura mater (Fig. 3d). The dura must be completely opened to the burr hole edges to allow passage of the endoscope without stripping the dura and causing an epidural hematoma. At this point the ROSA device's 3.2-mm-diameter steel bushing used for drilling is replaced with a shorter length 3.2-mm plastic bushing used to hold the endoscope. The shorter plastic bushing is advanced close to the entry site, about 5–10 mm from the scalp, so that the endoscope will reach its target depth. A ventricular catheter is passed through the plastic bushing and bur hole into the lateral ventricle to create our initial tract through the cerebral cortex. The ventricular catheter is then removed. The endoscope is passed through the bushing and follows the premade tract into the lateral ventricle where the intraventricular anatomy should be visually identified. For the purposes of endoscopy, we use the Aesculap Paediscope featuring a 3-mm outer diameter scope, with 2 working channels and a 150-mm length shaft (Fig. 3e).

The foramen of Monro is identified and should be in perfect alignment along the trajectory. The endoscope is then advanced into the third ventricle. The mammillary bodies are visualized through the transparent floor of the third ventricle as well as the infundibular recess more anteriorly. Using the "isometric" movement mode of the ROSA, minor manipulations of the endoscope position are done to direct the instrument channel at the exact spot chosen for fenestration along the floor of the third ventricle, just anterior to the mammillary bodies and basilar artery. The "isometric" movement mode restrains the movement of the endoscope at the predefined fulcrum point of the entry site which was previously defined at the burr hole through the skull. Adjustments of the endoscope trajectory are no more than 1–2 mm to optimize the instrument alignment for fenestration. This is done with the ROSA arm in the "slow" movement mode to avoid overcorrection. Once the exact alignment is achieved, the robotic arm is locked into position. A 1.0-mm-diameter Bugbee wire is passed through the working channel of the endoscope to puncture the floor of the third ventricle without electrocautery. Following this step a Fr-2 Fogarty balloon is inserted through the small perforation and the balloon inflated to dilate the fenestration to its full opening (Fig. 3f).

Once the fenestration window is complete, the endoscope is advanced through the opening created in order to visualize the basilar artery and ensure that the membrane of Liliequist has also been opened sufficiently with the fenestration (Fig. 3g). The endoscope is then gently backed out of the ventricular system while giving special attention to the structures surrounding the foramen of Monro to identify any signs of stretching or bruising of the fornices or the thalamic structures. Once the endoscope is removed from the patient's head, the ROSA's robotic arm is manually backed away from the patient and the skin incision is closed with a single 3-0 absorbable monofilament suture.

4.3 Post-operative Care

Following completion of the procedure and recovery from general anaesthesia, the patient is transferred from the recovery area to the surgical wards for overnight observation. In the absence of other medical indications, the patients are typically discharged the following day with planned follow-up in outpatient clinic the next week.

5   Tips and Tricks

1. Indications for ETV are key: Older patients with obstructive hydrocephalus never before shunted are the most optimal patients for ETV management of hydrocephalus. The ETVSS provides a valuable measurement tool to predict which patients might avoid VP shunt placement through fenestration of the third ventricle.

2. Contraindications:

   (a) Small children with open sutures and thin skulls pose a challenge for headframe application for stereotactic procedures in general.

   (b) Macrocrania from chronic hydrocephalus sometimes results in a trajectory length longer than our pediatric endoscope can reach (>12 cm).

   (c) The small burr hole for the robotic procedure does not allow for the extensive manipulation demanded for concomitant intraventricular procedures such as choroid plexus coagulation, septum pellucidum fenestration or biopsy of intraventricular tumors.

3. Opening the dura: The dura must be completely open to the burr hole edges to allow safe passage of the 3 mm outer diameter endoscope. We found that the #67 beaver blade scalpel fits best through the burr hole and has a straight edge to cut dura flush to the bone edge on each quadrant of the cruciate incision.

4. Isometric movements: The entry site of the trajectory should be defined at the mid-point of skull thickness to allow for the minor endoscope manipulations in "isometric" movement mode needed to optimize instrument position for ultimate fenestration.

6   Conclusions

The ROSA system provides a stable, precise, and minimally invasive approach to ETVs. Barriers to treatment caused by endoscope length and limits of endoscope manipulation related to burr hole diameter might be overcome in the future through adoption of longer endoscope shafts or larger diameter robotic bushings for drilling wider bur holes. Future studies are warranted to determine whether hydrocephalus outcomes might be improved and complications reduced through the adoption of this new technology. The robotic assisted ETV allows for minimal incision size requiring single stitch closure promoting rapid recoveries and high patient satisfaction.

References

1. Cohen A, Haines SJ (1995) Minimally invasive techniques in neurosurgery. Lippincott Williams & Wilkins, Philadelphia, PA

2. Hellwig D, Grotenhuis JA, Tirakotai W, Riegel T, Schulte DM, Bauer BL et al (2005) Endoscopic third ventriculostomy for obstructive hydrocephalus. Neurosurg Rev 28 (1):1–34. discussion 5-8

3. Kulkarni AV, Drake JM, Mallucci CL, Sgouros S, Roth J, Constantini S et al (2009) Endoscopic third ventriculostomy in the treatment of childhood hydrocephalus. J Pediatr 155(2):254–259.e1

4. Breimer GE, Sival DA, Brusse-Keizer MG, Hoving EW (2013) An external validation of the ETVSS for both short-term and long-term predictive adequacy in 104 pediatric patients. Childs Nerv Syst 29(8):1305–1311

5. Durnford AJ, Kirkham FJ, Mathad N, Sparrow OC (2011) Endoscopic third ventriculostomy in the treatment of childhood hydrocephalus: validation of a success score that predicts long-term outcome. J Neurosurg Pediatr 8 (5):489–493

6. Kulkarni AV, Riva-Cambrin J, Holubkov R, Browd SR, Cochrane DD, Drake JM et al (2016) Endoscopic third ventriculostomy in children: prospective, multicenter results from the hydrocephalus clinical research network. J Neurosurg Pediatr 18(4):423–429

7. Navarro R, Gil-Parra R, Reitman AJ, Olavarria G, Grant JA, Tomita T (2006) Endoscopic third ventriculostomy in children: early and late complications and their avoidance. Childs Nerv Syst 22(5):506–513

8. Schroeder HW, Niendorf WR, Gaab MR (2002) Complications of endoscopic third ventriculostomy. J Neurosurg 96(6):1032–1040

9. Erşahin Y, Arslan D (2008) Complications of endoscopic third ventriculostomy. Childs Nerv Syst 24(8):943–948

10. Hoshide R, Calayag M, Meltzer H, Levy ML, Gonda D (2017) Robot-assisted endoscopic third ventriculostomy: institutional experience in 9 patients. J Neurosurg Pediatr 20 (2):125–133

11. Lefranc M, Capel C, Pruvot-Occean AS, Fichten A, Desenclos C, Toussaint P et al (2015) Frameless robotic stereotactic biopsies: a consecutive series of 100 cases. J Neurosurg 122(2):342–352

12. Gonzalez-Martinez J, Mullin J, Vadera S, Bulacio J, Hughes G, Jones S et al (2014) Stereotactic placement of depth electrodes in medically intractable epilepsy. J Neurosurg 120 (3):639–644

13. Gonzalez-Martinez J, Vadera S, Mullin J, Enatsu R, Alexopoulos AV, Patwardhan R et al (2014) Robot-assisted stereotactic laser ablation in medically intractable epilepsy: operative technique. Neurosurgery 10(Suppl 2):167–172; discussion 72-3

14. Serletis D, Bulacio J, Bingaman W, Najm I, González-Martínez J (2014) The stereotactic approach for mapping epileptic networks: a prospective study of 200 patients. J Neurosurg 121(5):1239–1246

15. Lonjon N, Chan-Seng E, Costalat V, Bonnafoux B, Vassal M, Boetto J (2016) Robot-assisted spine surgery: feasibility study through a prospective case-matched analysis. Eur Spine J 25(3):947–955

Chapter 10

Robotic Automated Skull-Base Drilling

Bornali Kundu and William T. Couldwell

Abstract

The use of robots to aid in cranial surgery derives its origins from stereotactic surgery, which was later coupled with computed tomography imaging in the 1980s. The first robotic systems were navigational and performed procedures on deep brain structures such as brain biopsies and depth electrode placement. Since that time, autonomous and semiautonomous robots have been developed that use navigation coupled with guiding either surgical manipulators or surgical tools such as drills to aid with complex open skull-base surgery. The applications currently being explored include drilling of the temporal bone for resection of vestibular schwannomas, placement of cochlear implants, and performing mastoidectomies. These procedures are otherwise time consuming and require millimetric accuracy; however, because they can be largely done with image guidance and real-time tool position information feedback, they may be partially done by a robot. This is an ideal application of Computer Assisted Design–Computer Automated Manufacturing (CAD/CAM) principles because the bone is a rigid structure that will not deform significantly of shift during surgery. In this chapter, we discuss the current prototypes of robots available for aiding in skull-base surgery.

Key words Skull-base surgery, Translabyrinthine approach, Temporal bone, Drilling, Stereotactic navigation, Cranial

1 Introduction

Surgical robotics may have a role in making surgery more efficient because they are able to perform portions of an operation that are consistent from patient to patient, time consuming, and prone to human error. Robotic applications for cranial surgery fall into several categories: those focused on navigation and operations done on deep brain structures (e.g., the ROSA robot is used for placement of depth electrodes) [1], those using robotic arms to perform surgical procedures (e.g., the Da Vinci robot is used to assist in microsurgery or endoscopic skull base procedures [2] or the NeuroArm, which has arms and uses the intraoperative

Electronic supplementary material: The online version of this chapter (https://doi.org/10.1007/978-1-0716-0993-4_10) contains supplementary material, which is available to authorized users.

Hani J. Marcus and Christopher J. Payne (eds.), *Neurosurgical Robotics*, Neuromethods, vol. 162, https://doi.org/10.1007/978-1-0716-0993-4_10, © Springer Science+Business Media, LLC, part of Springer Nature 2021

magnetic resonance imaging suite [3]), and lastly those focused on performing more accurate, efficient drilling for open skull-based procedures [4–6]. In this chapter, we focus on this last category of robots.

Both semiautonomous (i.e., requiring human interaction) and autonomous systems have been built. Autonomous milling systems are ideal for skull-base applications because parts of the operation involve drilling bony anatomy, which is rigid and thus consistent during the operation. This feature allows the surgeon to formulate a presurgical plan to allow a robot to perform a portion of the surgery. Autonomous milling systems have been explored for applications such as drilling the temporal bone for performing a translabyrinthine approach to resect a vestibular schwannoma or perform a mastoidectomy [4–6]. There may be applications where drilling near the sella can be done via a transnasal or transoral approach since the bony anatomy is consistent intraoperatively [7]. Procedures where an automated system would not be ideal are those performed under conditions where the anatomy can change with time, such as resecting tissue that can hemorrhage or change in shape during the operation.

Using a skull base surgery robot to perform time-consuming drilling has great appeal. There is potential to reduce cost by shortening the operating time, lower the risk of infection by shortening the time the wound is open, and decrease surgeon fatigue [4]. However, robots can be prohibitively costly and not as efficient as advertised, which has tempered a surge in their use overall [8]. Nevertheless, there is certainly an increasing interest in the field, as demonstrated by the increasing number of publications on the topic in the last decade [8].

2   Summary of Evidence

There are several reported prototype robots for skull-base work. None of these systems have yet been approved by the U.S. Food and Drug Administration for patient use. The evidence supporting their use comes from cadaveric demonstrations of their performance. The versatility of the robot to be able to perform a drilling plan depends on the number of degrees of freedom the robotic system has. For example, the ROSA robot, which is used for navigating depth electrodes and is in wide commercial use, has 6 degrees of freedom [1]. The accuracy of the drilling is largely defined by the accuracy of the navigation system guiding the robot. The largest part of this is patient registration to his or her imaging, a process that is defined by fiducial localization. Bony fiducials are the most robust and offer the best accuracy, minimizing fiducial localization error [9].

Patient-specific anatomy is used for the current skull-base drilling robot prototypes. Segmentation algorithms use the imaging data to calculate a model surface of the bone anatomy. Within that, the surgeon can define the volume of bone to be drilled. Beyond this rigid anatomy, features of bone density and porosity derived from imaging data may be helpful to define the drilling force and rotations per minute (RPMs) used for sections of bone near critical structures [10]. RPMs can be guided by real-time feedback from sensor data such as electromyography of, e.g., facial muscles in the case of facial nerve monitoring [5, 11]. It may be possible to optimize solutions to the drilling plan based on prior data from pre- and postoperative imaging of surgeries done by expert surgeons. Specifically, the robotic solution can be validated by its ability to reproduce the surgeon's solution. This information may even be used to teach an adaptive system to learn the best drilling path based on data from an experienced surgeon.

With any new technology, a period of assessment must elapse during which the technology is evaluated in laboratory, cadaveric, and clinical environments. The IDEAL guidelines offer recommendations for study design and reporting standards of this work [12]. The guidelines offer a parallel for medical devices and procedures to be judged to the way new medicines are approved. The current skull base prototype robots are in the cadaveric testing stages (IDEAL Stage 0).

3   Description of Robot

There are three prototype robots reported in the literature at the time of this writing for drilling the skull base [4–6]. We focus here on the prototype described by Couldwell et al. [4]. The computer assisted design/computer automated manufacturing (CAD/-CAM) skull base drill (Fig. 1) integrates image-guided planning with real-time feedback sensor information to achieve a drilled result. It can be used for both cranial and spine operations. It uses patient specific anatomy derived from a stereotactic resolution (less than 1 mm × 1 mm × 1 mm) computed tomography imaging to define the drilling plan. The surgeon defines the volume of drilling on the two or three-dimensional reconstruction of the anatomy (Fig. 2). The system will then compute the optimal drilling path to achieve the volume. The software algorithm is a robust computer numerical controlled (CNC) surgical cutting system that interfaces with 3D medical imaging data [13]. This CNC algorithm is used in a variety of manufacturing settings, particularly for milling and shaping parts, and is integrated with the CAD and CAM software.

The robot itself has 5 degrees of freedom executed on a multi-axis kinematic system via a rigid mechanical design (x, y, z, roll, pitch; Fig. 1). This allows for a large range of drill angles to be

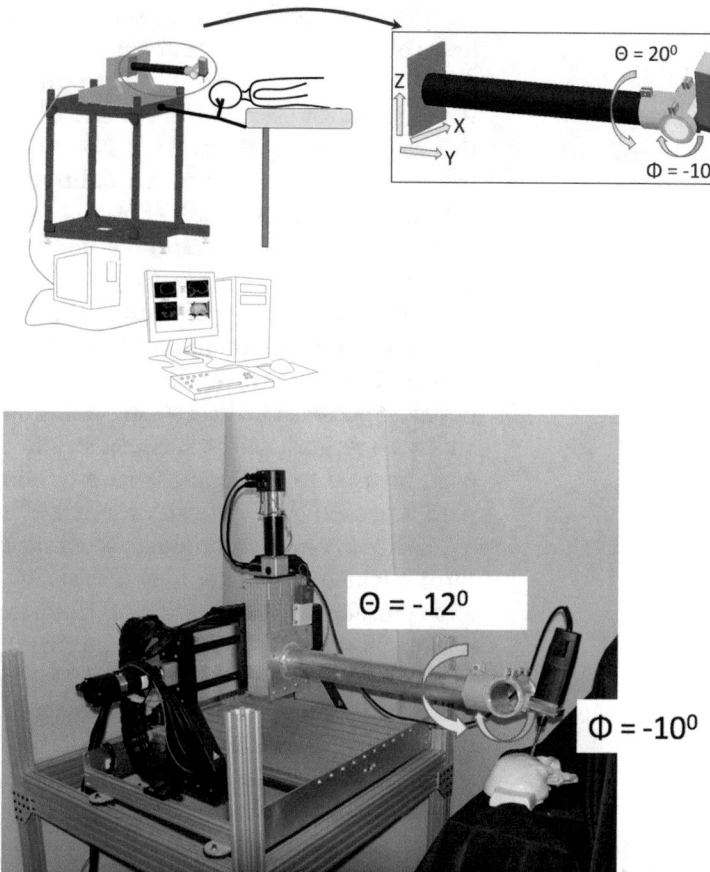

Fig. 1 (Top) Schematic demonstrating the software interface with the surgical machining system (inset enlarged). (Bottom) Photograph showing the robot with the attached surgical arm and drill. (Reproduced with permission from Couldwell WT, MacDonald JD, Thomas CL, Hansen BC, Lapalikar A, Thakkar B, Balaji AK (2017) Computer-aided design/computer-aided manufacturing skull base drill. Neurosurg Focus 42:E6)

achieved by the machine to create the surgical plan. Safety features include real-time monitoring of electromyographic data, e.g., from the facial muscles signaling proximity to the facial nerve in a translabyrinthine approach. The robot also uses anatomical data to define regions of avoidance such as the venous sinuses (Fig. 3) and the facial canal (Figs. 4 and 5). The objective is to drill most of the bone quickly, leaving a margin of 1 mm of bone around vital structures. In cadaveric studies, this drill can perform gross removal of bone for a translabyrinthine approach in 2½ min. This is something that takes an expert surgeon approximately 2 h to perform.

In contrast, the prototype developed by Lim et al. [6] is semi-autonomous such that the surgeon must physically guide the drill during the procedure. The drill does not have the weight of gravity,

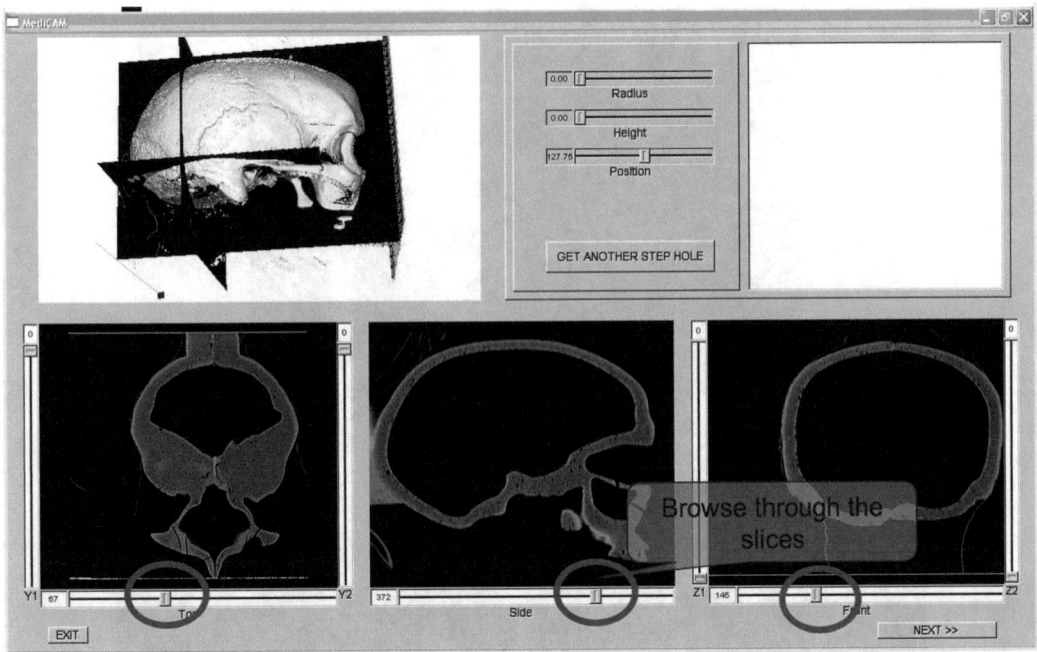

Fig. 2 Three-dimensional model of the bony anatomy is displayed in a graphic interface so the surgeon can indicate the portion of bone or soft tissue to be machined. (Reproduced with permission from Couldwell WT, MacDonald JD, Thomas CL, Hansen BC, Lapalikar A, Thakkar B, Balaji AK (2017) Computer-aided design/computer-aided manufacturing skull base drill. Neurosurg Focus 42:E6)

which may be unusual for a trained surgeon to acclimate to. This drill also has 5 degrees of freedom and has been tested to leave a margin of 2.5 mm between critical structures and the drill path before the safety shutoff is engaged. In a cadaveric study, engineers not trained in surgery were able to drill the temporal bone funnel and not injury critical structures within the 'segmented area' defined by the surgeon for resection. They did not report the time it took for perform the drilling.

Another prototype is the robot described by Dillon et al. [5, 10]. This robot is also autonomous, with 4 degrees of freedom (x, y, z, and roll) and uses similar bony features to guide drilling and similar safety features as that of Couldwell et al. The algorithm also incorporates features such as angle of the drill and shape of the burr defining the forces delivered to the tissue. The estimated time to drill the temporal bone funnel for a vestibular schwannoma resection was 22–26 min and the actual distance between drill path and the facial nerve was 0.5 mm on average (which was what was planned). Thus, more time is needed to program the drill to mill closer to vital structures as would be expected. This machine also uses a CNC-based algorithm.

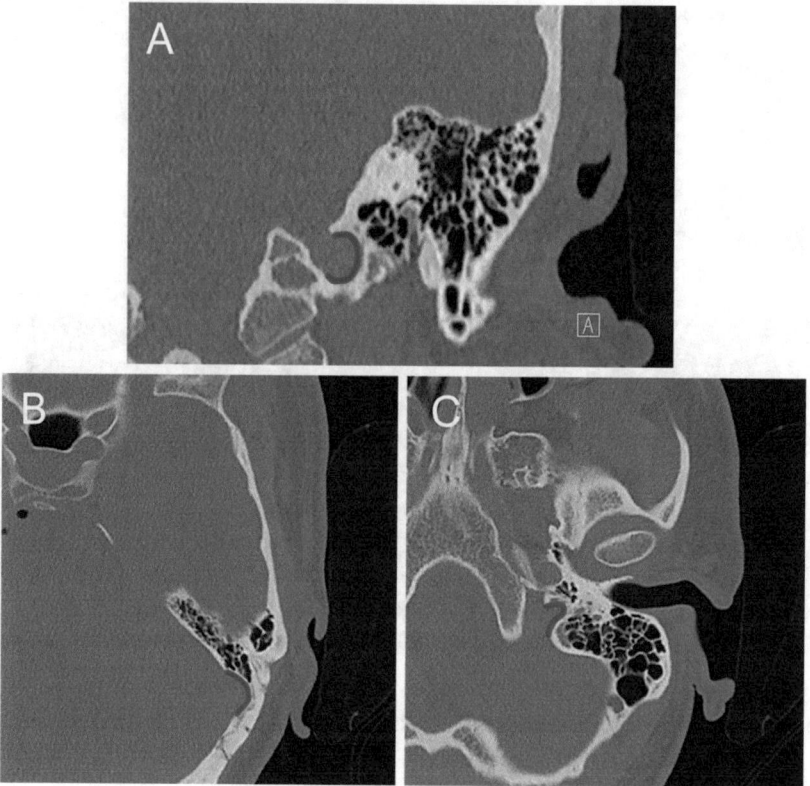

Fig. 3 CT left temporal bone demonstrating the outline of the sigmoid sinus and jugular bulb in coronal (**a**) and axial (**b**, **c**) planes. The sinus is outlined to develop a region of avoidance for the drill. (Reproduced with permission from Couldwell WT, MacDonald JD, Thomas CL, Hansen BC, Lapalikar A, Thakkar B, Balaji AK (2017) Computer-aided design/computer-aided manufacturing skull base drill. Neurosurg Focus 42:E6)

4   Description of Technique

For the CAD/CAM skull base drill described by Couldwell et al. [4], surgical planning is considered during the CAD phase. In this phase, medical imaging is used to create the computer model of the (bone) volume to be operated upon. There can be integration of multiple imaging modalities such as magnetic resonance imaging at this point. For any plan, the surgeon can specify the features of the drilled bone volume including axis, radius, and depth within the CAD interface (Fig. 2). The surgeon can also define regions of avoidance such as the facial nerve. The system will then compute the optimized drilling path to achieve the result. We describe the technique in the application of temporal bone drilling for a transla-byrinthine approach to resect a vestibular schwannoma. The video of this procedure done on a cadaver is published in Couldwell et al. [4]. In this case, the drilled volume is a complex funnel with borders defined by the facial nerve and the sinuses (Figs. 3, 4, and 5).

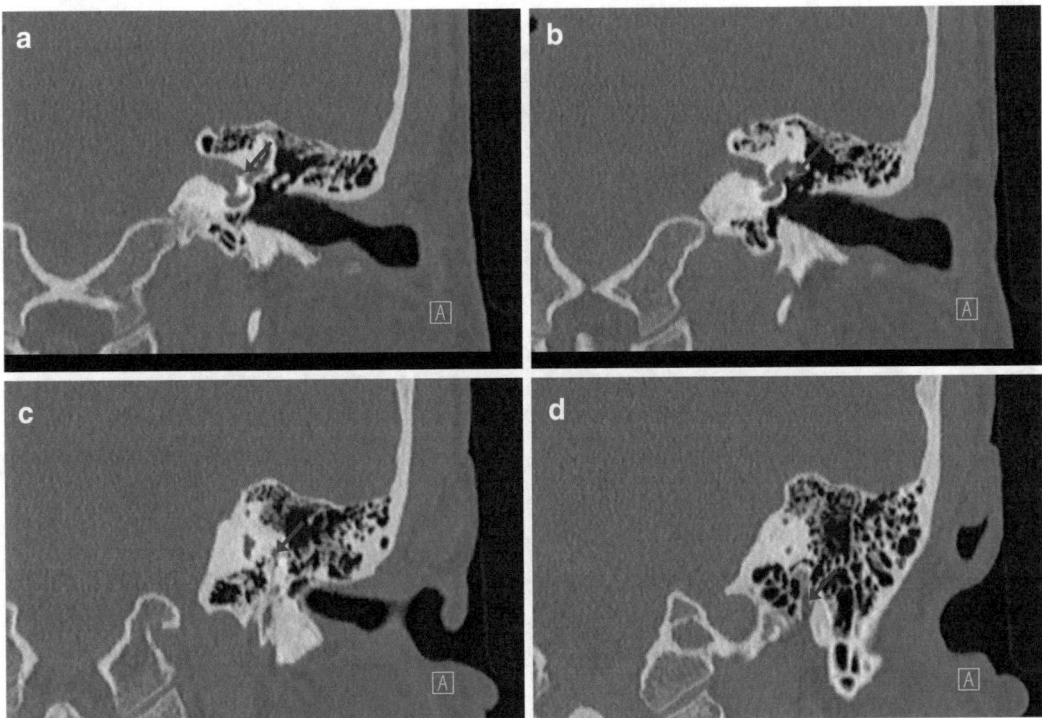

Fig. 4 CT left temporal bone coronal images (**a**–**d**). The arrow demonstrates the course of the facial nerve as it exits the internal auditory canal (**a**), through its course in the temporal bone (**b**–**d**). This would be a region of avoidance for the drill path. (Reproduced with permission from Couldwell WT, MacDonald JD, Thomas CL, Hansen BC, Lapalikar A, Thakkar B, Balaji AK (2017) Computer-aided design/computer-aided manufacturing skull base drill. Neurosurg Focus 42:E6)

The next phase is registration such that the CAD plan can be implemented in the CAM system. The robotic system is fixed to the rigid head holder holding the patient. The patient is then registered to his or her imaging with infrared-based frameless stereotaxy. As discussed above, registration error can significantly affect robot performance, but it is best to have this error as consistent as possible, since surgeons can generally compensate within their plan based on this known error.

Real-time control allows for emergency shutoff at any time as well as shutoff if irritation of the facial nerve is detected during the operation. Shutoff initially occurs if there is increased muscle activity above baseline for this prototype. The drill speed is high, as with current hand drills, to ensure smooth motion and reduce kickback. Once the robot has drilled most of the bone quickly, the surgeon can then finish drilling around sensitive vital structures (e.g., unroofing the facial nerve canal).

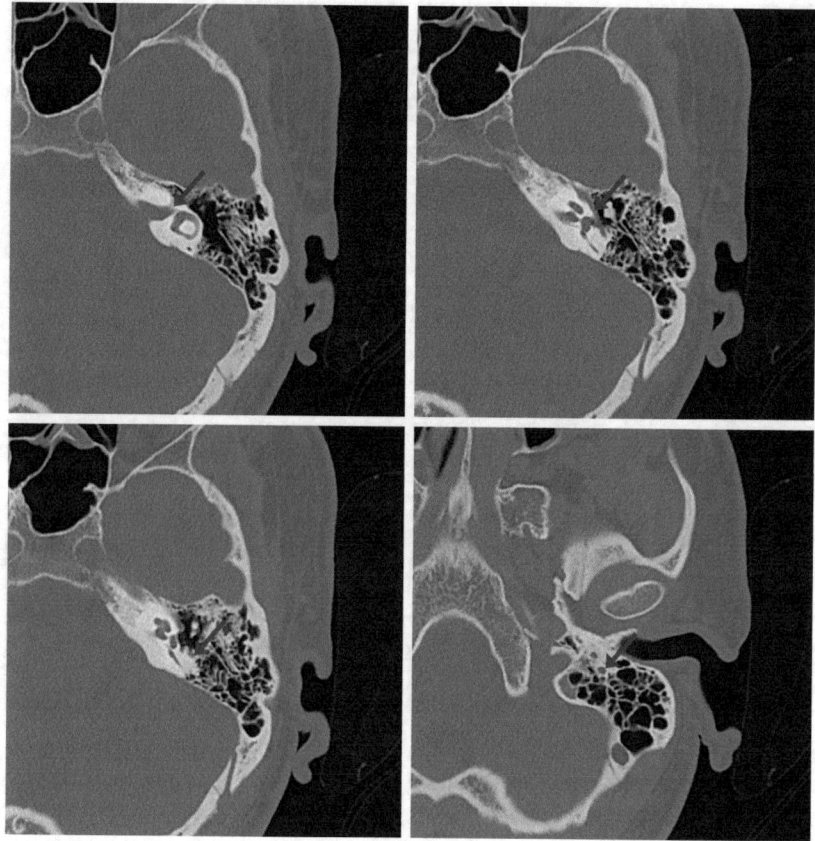

Fig. 5 CT temporal left bone coronal images (**a–d**). The red arrow outlines the course of the facial nerve through the temporal bone to its exit at the stylomastoid foramen. The software is designed to develop a region of avoidance for the drill. (Reproduced with permission from Couldwell WT, MacDonald JD, Thomas CL, Hansen BC, Lapalikar A, Thakkar B, Balaji AK (2017) Computer-aided design/computer-aided manufacturing skull base drill. Neurosurg Focus 42:E6)

5   Tips and Tricks

Skull base drilling robots are not currently in commercial use. Research to emulate the actions and decision trees an experienced surgeon performs and considers is underway [10]. The prospect of artificial intelligence learning from human performance may be possible. For example, it may be possible to incorporate the spatial pattern, temporal pattern, and force pattern of drilling that an expert surgeon uses and teach the software to use features of that pattern to optimize its solution.

6   Conclusions

Robot-assisted surgery may speed up skull-base surgery. It may allow for decreased operating room time, which is known to relate to cost, rates of infection, and outcomes. The key is to keep the machine itself small and usable and create algorithms that calculate the optimal drill trajectory in a short amount of time. The registration to the patient needs to be accurate and robust. Limitations of the current prototypes are the learning curve to set up the robot and ensure accurate registration. Anomalies in bone anatomy may also preclude some patients as candidates for robotic milling [5]. Changing burr size is also a challenge but may be necessary to optimize the drill path solution. These possibilities are being explored by these research groups.

7   Electronic Supplementary Material

Video S1 Robotic automated skull-base drilling of temporal bone in a cadaver (MP4 70,656 kb).

References

1. Fomenko A, Serletis D (2017) Robotic stereotaxy in cranial neurosurgery: a qualitative systematic review. Neurosurgery 83:642–650. https://doi.org/10.1093/neuros/nyx576

2. Hong W-C, Tsai J-C, Chang SD et al (2013) Robotic skull base surgery via supraorbital keyhole approach. Neurosurgery 72:A33–A38. https://doi.org/10.1227/NEU.0b013e318270d9de

3. Sutherland GR, Wolfsberger S, Lama S et al (2013) The evolution of neuroArm. Neurosurgery 72(Suppl 1):27–32. https://doi.org/10.1227/NEU.0b013e318270da19

4. Couldwell WT, MacDonald JD, Thomas CL et al (2017) Computer-aided design/computer-aided manufacturing skull base drill. Neurosurg Focus 42(5):E6. https://doi.org/10.3171/2017.2.focus16561

5. Dillon NP, Balachandran R, Siebold MA et al (2017) Cadaveric testing of robot-assisted access to the internal auditory canal for vestibular schwannoma removal. Otol Neurotol 38:441–447. https://doi.org/10.1097/MAO.0000000000001324

6. Lim H, Matsumoto N, Cho B et al (2016) Semi-manual mastoidectomy assisted by human-robot collaborative control - a temporal bone replica study. Auris Nasus Larynx 43:161–165. https://doi.org/10.1016/j.anl.2015.08.008

7. Chauvet D, Missistrano A, Hivelin M et al (2014) Transoral robotic-assisted skull base surgery to approach the sella turcica: cadaveric study. Neurosurg Rev 37:609–617. https://doi.org/10.1007/s10143-014-0553-7

8. Smith JA, Jivraj J, Wong R et al (2016) 30 years of neurosurgical robots: review and trends for manipulators and associated navigational systems. Ann Biomed Eng 44:836–846. https://doi.org/10.1007/s10439-015-1475-4

9. Gerber N, Gavaghan KA, Bell BJ et al (2013) High-accuracy patient-to-image registration for the facilitation of image-guided robotic microsurgery on the head. IEEE Trans Biomed Eng 60:960–968. https://doi.org/10.1109/TBME.2013.2241063

10. Dillon NP, Fichera L, Wellborn PS et al (2016) Making robots mill bone more like human surgeons: using bone density and anatomic information to mill safely and efficiently. Rep U S 2016:1837–1843. https://doi.org/10.1109/IROS.2016.7759292

11. Dillon NP, Fichera L, Kesler K et al (2017) Pre-operative screening and manual drilling strategies to reduce the risk of thermal injury during minimally invasive cochlear implantation surgery. Ann Biomed Eng 45:2184–2195. https://doi.org/10.1007/s10439-017-1854-0

12. McCulloch P, Altman DG, Campbell WB et al (2009) No surgical innovation without evaluation: the IDEAL recommendations. Lancet 374:1105–1112. https://doi.org/10.1016/S0140-6736(09)61116-8

13. Burger T, Laible U, Pritschow G (2001) Design and test of a safe numerical control for robotic surgery. CIRP Ann Manuf Technol 50:295–298

Chapter 11

Robot-Assisted Pituitary Surgery

Dorian Chauvet and Stéphane Hans

Abstract

The transsphenoidal approach represents the gold standard for surgical excision of pituitary tumors. Over the last few decades the endoscopic endonasal transsphenoidal approach has become widely accepted, with angled-endoscopes providing excellent visualization of the operative field and possibly allowing for a greater likelihood of complete resection. On the other hand, there are limitations of this technique, especially concerning potential rhinological side effects and the 2D vision offered by standard endoscopes.

Transoral robotic surgery (TORS) with the da Vinci system is now commonly used in head and neck cancers, and has been demonstrated to be safe and effective. Here, we describe an innovative TORS approach to the sella to remove pituitary tumors, from cadaveric studies to clinical proof of concept.

Key words Transoral robotic surgery, Pituitary adenoma, Skull base surgery, Robotic assisted surgery, Da Vinci system, Transsphenoidal surgery, Surgical technique

1 Introduction

The da Vinci system (Intuitive Surgical Inc., Sunnyvale, CA, USA) has increasingly undergone worldwide adoption as a result of its advantages: tremor filtration, 3D vision, motion scaling and great dexterity. It was designed for general surgery and its robotic technology is mostly used in the fields of urology and gynecology. Recently head and neck surgeons have applied robotic surgery to cancers of the pharynx and larynx [1, 2], and have introduced the term TransOral Robotic Surgery (TORS).

Within neurosurgery, two types of cadaveric works have been described prior to our clinical skull base project: transoral odontoidectomies [3, 4] and keyhole transcranial approaches [5]. For the latter, the conclusions were that the da Vinci system does not allow deep dissection in narrow corridors and does not offer haptic feedback.

From Crockard's adage *"transoral approach allows access to structures from the sphenoid sinus rostrally to the fourth cervical vertebral body caudally"* [6], we envisioned the possibility of

Hani J. Marcus and Christopher J. Payne (eds.), *Neurosurgical Robotics*, Neuromethods, vol. 162,
https://doi.org/10.1007/978-1-0716-0993-4_11, © Springer Science+Business Media, LLC, part of Springer Nature 2021

reaching the sella with an innovative transoral robotic approach using the da Vinci. To perform it, we placed the 30° endoscope behind the posterior edge of the hard palate, looking upwards. From cadaveric preliminary works [7] and anatomical studies [8] to clinical first proof of concept [9], we present herein this innovative TORS for pituitary tumours.

2   Summary of Evidence

Skull base pathologies include sellar tumours such as pituitary adenomas. These tumours were initially managed using a supratentorial approach, but transphenoidal surgery is now the gold standard due to absence of brain retraction. Most neurosurgical teams use an endoscopic endonasal technique because it allows for greater visualization of the sphenoid sinus and sella [10]. Moreover, endoscopic techniques also allow for more extensive skull base surgery when required through the so called expanded endoscopic endonasal approach [11]. However, there are limitations of the endoscopic endonasal technique especially concerning potential rhinological side effects and the 2D vision offered by standard endoscopes.

The da Vinci system offers a 3D endoscope and the possibility to perform microsurgical dissection with increased dexterity. We therefore hypothesized that robotic assistance could be of benefit in anterior skull base surgery.

Our project was carried in three sequential stages: preliminary cadaveric dissections, anatomical study and clinical first protocol.

2.1   Preliminary Cadaveric Dissections [7]

We performed 11 dissections on fresh human cadavers to explore the accessibility of the sella via TORS and to optimize the technique (described in the next section). In all cases, the visualization of the cavum, the mucosal dissection, the sphenoid drilling (*see* Fig. 1) and the opening of the sella were satisfactory, despite the anatomical heterogeneities of the specimens. The robotic instruments could be inserted into the sphenoid sinus in all cases and it also reached the sella turcica in all procedures with acceptable manoeuvrability. Then a virtual tumour removal was performed on the pituitary gland (*see* Fig. 2). We did not experience lateral deviation in our dissections.

From lateral fluoroscopic views, we measured the angle of work to reach the skull base: (1) when the drill was placed at the midline and (2) when it was placed in the labial commissure of mouth. Indeed, we hypothesized that the lateral movement of the handpiece, following naturally the lower teeth curve from the midline to the labial commissure, allowed an opening of the angle of work to the skull base (*see* Fig. 3). The latter was defined by the angle between the horizontal line passing through the hard palate and

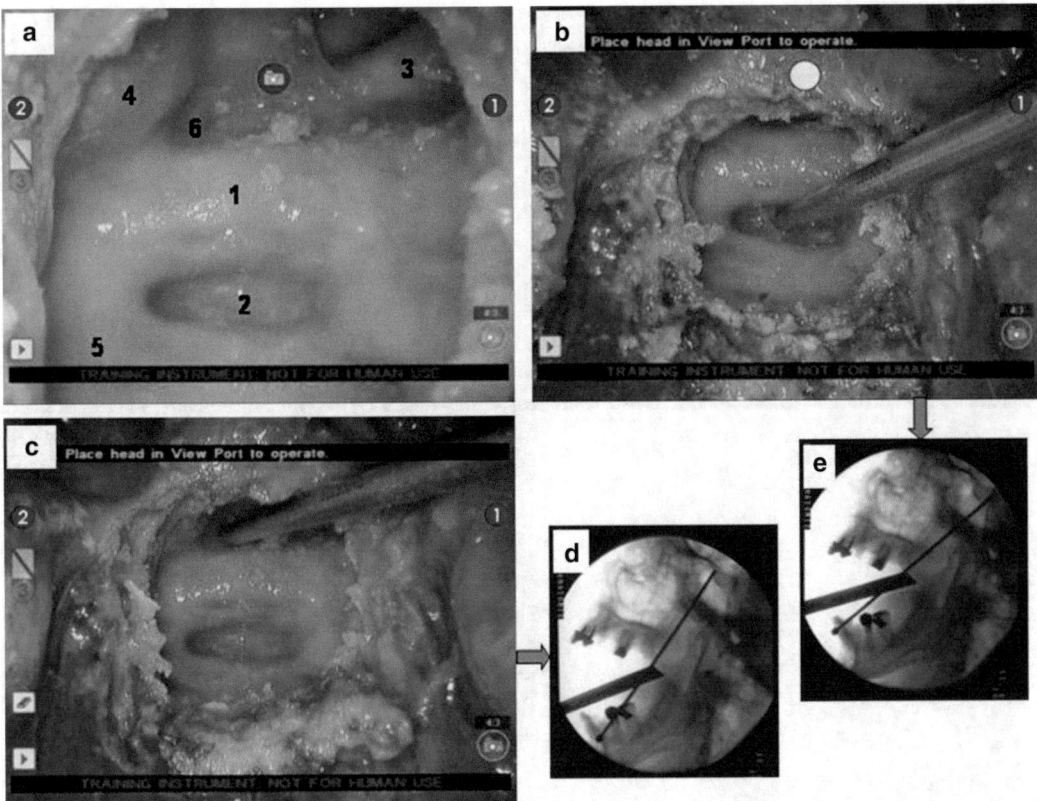

Fig. 1 Intraoperative endoscopic view of the sella turcica. (**a**) anatomical structures of the sphenoid sinus: (1) sellar floor, (2) dorsum sellae that is well pneumatized, (3) left optic nerve protuberance, (4) right carotid protuberance—sellar portion, (5) right carotid protuberance—clival portion and (6) opto-carotid recess. (**b**) Dissector placed in the pneumatized dorsum sellae and the corresponding fluoroscopic lateral picture (**e**). (**c**) Dissector inserted in front of the anterior wall of the sella and the corresponding fluoroscopic view (**d**) [7]

the projected line of the drill. From these cadaveric data, we observed that the mean angles of work were 55° (min 48, max 62) and 71.5° (min 67, max 76) for the midline and the lateral positions respectively. Hence we demonstrated that the mean angle of work gained by placing the drill at the labial commissure was +16.5°.

In conclusion these preliminary results were satisfactory and TORS skull base could be envisioned for selected patients.

2.2 Anatomical Study [8]

Before starting the clinical study, we conducted a prospective radiological study with the hypothesis that TORS for skull base would be feasible in the majority of patients, regardless of their anatomical features. Hence 30 consecutive patients with neurological issues and without past medical history of endonasal surgery, sinus disease and/or skull base pathology were included. We assessed the following anatomical criteria on open-mouth cerebral CT-scan, bony window sagittal midline view (*see* Fig. 4):

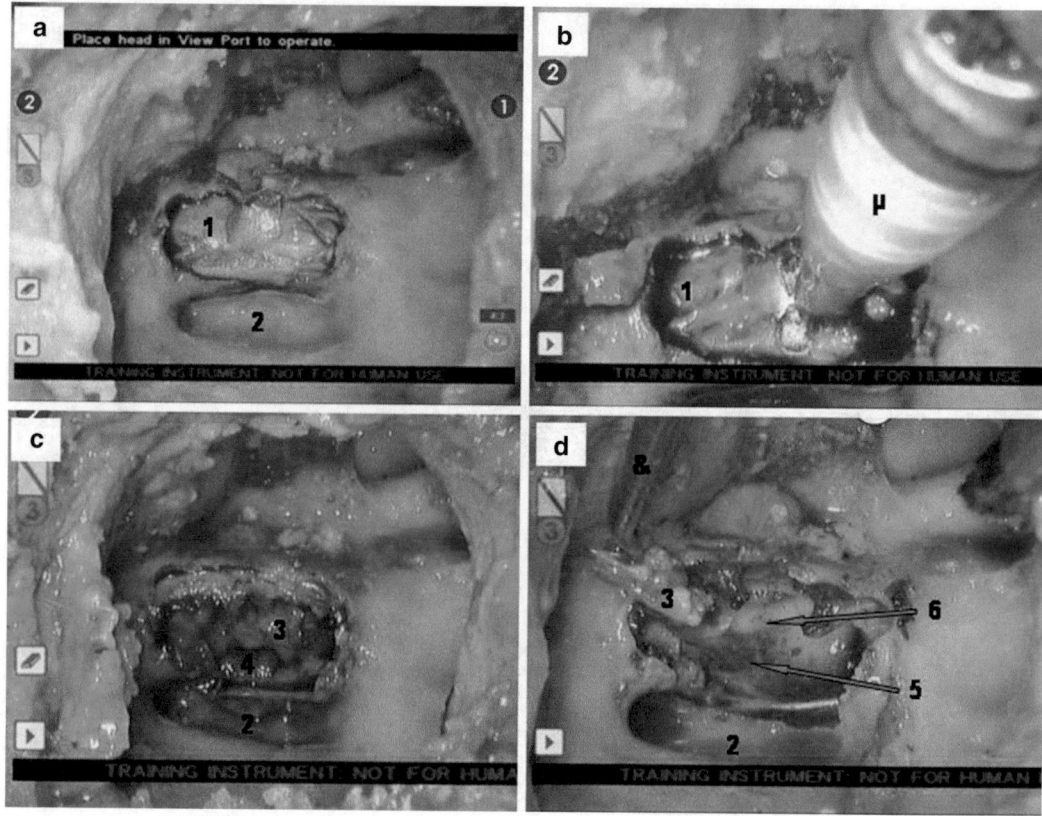

Fig. 2 Intraoperative view of the pituitary fossa dissection. (**a**) View after sellar floor removal. (**b**) Cauterization of the sellar dura with the monopolar cautery (μ). (**c**) View during pituitary gland resection. (**d**) Final view after removal. Legends: (1) sellar dura, (2) pneumatized dorsum sellae, (3) pituitary gland, (4) sellar diaphragm, (5) pituitary stalk retracted by a hook (&) and (7) optic chiasm [7]

- the lowest point of the sella turcica, point A,
- the most posterior palatine bone landmark, point B,
- the most anterior palatine bone landmark, point C,
- the maxillary dental point, at the tip of superior incisor, point D,
- the mandibular dental point, at the tip of inferior incisor, point E.

Then we observed the (BE) line projection, which corresponded to the dissection axis, in a midline plane and in a 25° oblique plane corresponding to the intraoperative position of the instruments (*see* Fig. 5). For both situations, patients were classified in three categories: projection anterior to the point A of the sella (aka presellar projection), on the point A and posterior to the point A (aka postsellar projection) (*see* Fig. 4b, a and c respectively). We found that 40% of patients ($n = 12$) had a (BE) line projection that moved forward from the projection on the A point to the presellar projection, when studied in an oblique plane.

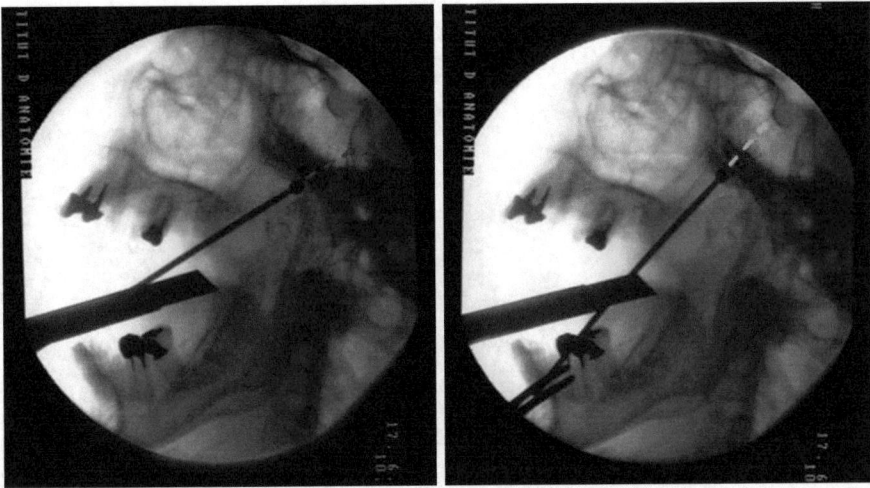

Fig. 3 Fluoroscopic lateral views, the endoscope standing at the midline of the mouth. On the left, the matchstick drill is inserted at the midline and its projection on the sphenoid bone virtually meets the clivus (red dotted line). On the contrary on the right picture, the bur is placed in the labial commissure and its projection clearly meets the sella turcica (green dotted line). This shows how the angle of work to the skull base is increased when the instruments are placed laterally in the oral cavity [8]

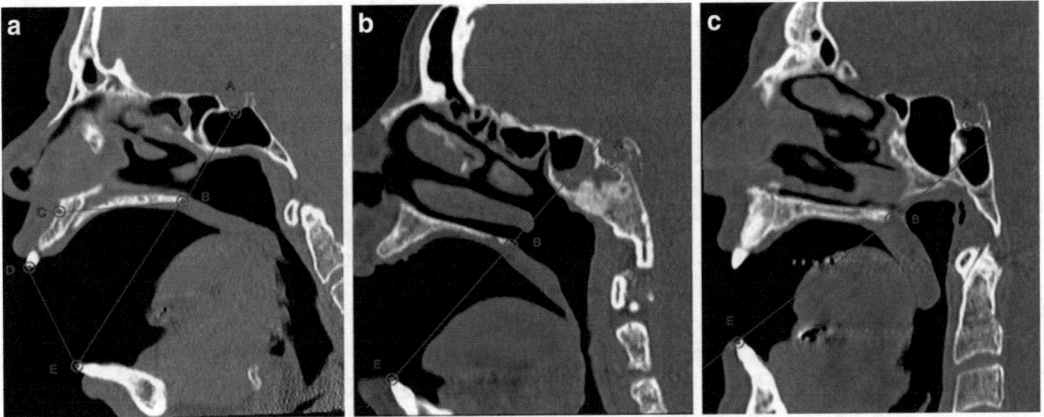

Fig. 4 CT scan bony window sagittal midline views. (**a**) Description of the different points: A sella turcica point, B posterior palatine point, C anterior palatine point, D maxillary dental point, E mandibulary point. Notice that the (BE) line is projected on the point A on this picture. (**b**) Presellar projection of (BE) line. (**c**) Postsellar projection of (BE) line [8]

As a predictive factor of surgical feasibility, we chose the projection of the (BE) line on the sella and patients were divided in two groups: the "straight approach" group with a presellar projection and the "no straight approach" group (with projection on the point A of the sella and postsellar projection). The latter was considered as unfavourable for TORS. With statistical analysis—Student's t-test—we observed that the only significant predictive factor was the spontaneous mouth opening [DE] ($p < 0.05$). We also

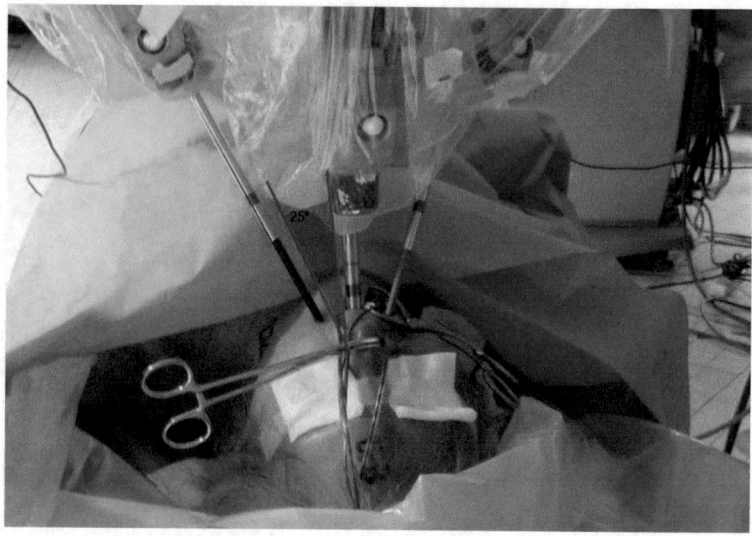

Fig. 5 Photograph from cadaveric dissection of TORS at the bedside of the cadaver. The robotic instruments are placed into the oral cavity with a 25° angle (represented with a red triangle) between the endoscope at the midline and the dissector laterally [8]

demonstrated that a mouth opening of 38.9 mm may be large enough to obtain a sensitivity of 83% and a specificity of 70.8% to predict our straight approach hypothesis.

In conclusion, these anatomical data emphasized that the physiological maximal mouth opening could be a good predictive factor for feasibility of TORS, knowing that the mouth aperture could be larger intraoperatively (with retractor) than the spontaneous one. Thereby it seemed evident that patients with trismus could not be included in a clinical study.

2.3 Clinical First Protocol [9]

This single-center prospective study aimed to confirm the accessibility of the sella with TORS. It was conducted following French ethical committee validation and Clinical Trials registration (NCT02743442). A total of seven patients were included (5 female, 2 male; mean age 46 years old). All presented with bitemporal heminanopia, except one patient who had a sellar tumor growth on serial imaging.

Regarding radiological preoperative data, 5 tumours were partially or totally cystic and 2 were totally solid (*see* Fig. 6). All tumours except one had a suprasellar extension that was responsible of the visual field defect. Mean size of the tumors' largest dimension was 29 mm (min 21 – max 39). Open mouth CT-scan demonstrated that all patients had a well pneumatized sphenoid sinus and that the projection line on the sella (described above) were divided into 2 presellar, 3 sellar and 2 postsellar.

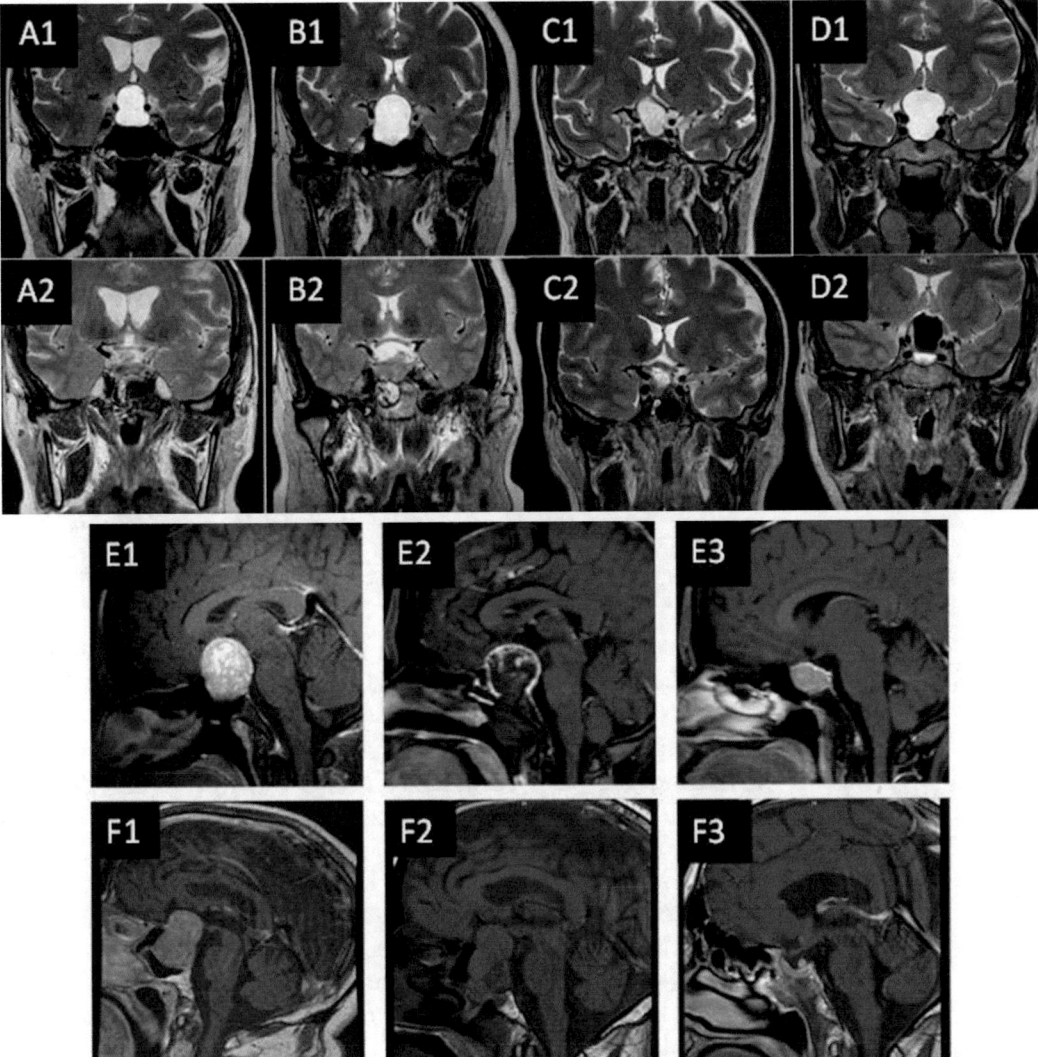

Fig. 6 Upper figure shows coronal T2-weighted brain MR imagings, preoperatively (1) and postoperatively at day 1 (2). A, B, C and D for patients n°1, 2, 3 and 4 respectively. For patient n°4, postoperative imaging (D2) shows intrasellar hyposignal corresponding to pneumocephalus [9]. Lower figure shows patient n°5 and 6, respectively E and F. E1 peroperatively, E2 postoperatively at day 1 with hematoma within the sella, E3 postoperatively at 1 month with a resorption of the hematoma and the chiasmatic decompression. F1 preoperatively, F2 postoperatively at day 1 with a partial reduction of the tumour, F3 postoperatively after second endonasal surgery

Intraoperatively, the visualization of the cavum was satisfactory in all cases. Nevertheless we reported one patient whose operative corridor was narrowed because of thick tonsils and mucosa. The visualization of the sella was also good and the opening of the sella allowed access to the tumour in all cases. Afterwards the removal quality depended on the tumor consistency. If the tumour was cystic ($n = 5$), the fluid drained easily. If the tumor was totally

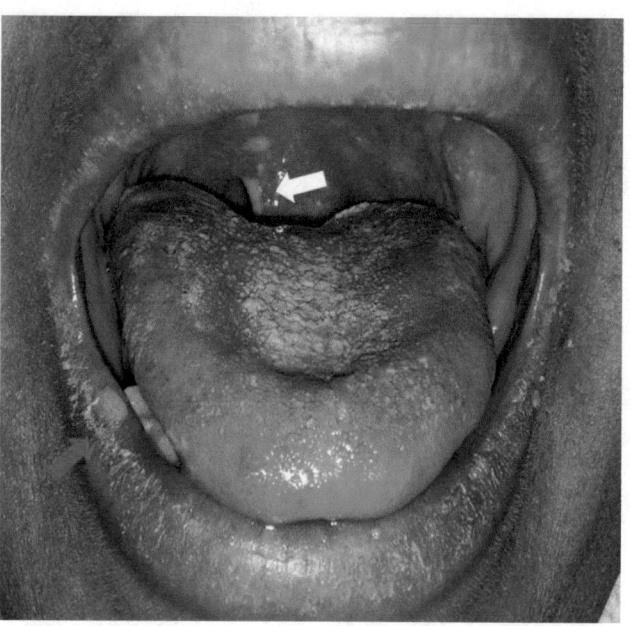

Fig. 7 Postoperative day 7 photograph of patient n°2. Blue arrow and white arrow point out the lower lip and soft palate lesions respectively

solid ($n = 2$), the removal was hard due to two issues: (1) the dissection was very haemorrhagic and we hypothesize that the position of the head (slightly extended) was a possible reason; (2) the da Vinci system has no dedicated instruments such as curettes. These issues led to partial removal and one patient had to be reoperated on via endonasal approach (*see* Fig. 6).

At the end of the procedures, we observed 3 minor lower lip lesions (because of the instruments at the labial commissure) and 2 minor mucosal lesions next to the uvula (because of the loops that retracted the soft palate) (*see* Fig. 7).

At 1 month after surgery, all patients experienced improved vision. No rhinologic disturbance was reported. We observed the following TORS side effects: sore throat ($n = 7$), hypernasal speech ($n = 5$) and otitis media ($n = 1$). Fortunately these symptoms were transient (approximately 3–5 days). We also reported the following complications of sellar surgery: CSF leak ($n = 1$, resolved after lumbar puncture), diabetes insipidus ($n = 2$) and hypopituitarism ($n = 1$).

Regarding the postoperative MRI, we had good results about cystic lesions, but the removal quality was poor for the two solid tumors (*see* Fig. 6).

Finally, we reported three pathological confirmations of gonadotroph adenomas. The other lesions were mainly cystic without diagnoses.

3   Description of Robot

The robot used in our skull base project is a teleoperated system: this means that the surgeon performs a remote dissection, placing both hands in dedicated joysticks that command the robotic arms. This telesurgery was initially created to operate on astronauts during mission [12]. Then military programs developed the concept [13].

The da Vinci system (Intuitive Surgical Inc., Sunnyvale, CA, USA) is a teleoperated robot. Several advantages have been emphasized. The camera arm has a double video endoscope device that provides clear 3D vision for the surgeon at the console. The system offers a motion scaling, a tremor filtration and increased dexterity within narrow spaces. Additionally the ergonomic comfort for the surgeon has been published [3]. The system includes a surgical console with master controllers, a patient cart with robotic arms and a vision cart (*see* Fig. 8).

Concerning our project, we initially used for the cadaveric work a S HD 4 arms da Vinci system, located at the Ecole Européenne de Chirurgie, Paris. Later during the clinical study, we perform the surgery on a SI HD 4 arms system. In both procedures, we only used 3 arms (see below): the videoendoscope, a 5 mm EndoWrist® Maryland dissector and a 5 mm EndoWrist® Monopolar cautery.

The operative room setup is the first decisive matter, as the robotic device takes much space. From our preliminary work, we propose the following setup to perform skull base TORS (*see* Fig. 9).

4   Description of Technique

First we place a mouth retractor (type Doyen, Landanger®) to obtain the best transoral exposition. We retract the soft palate with two rubber catheters introduced into the nose and pulled out by the mouth (*see* Fig. 10). We insert the robotic arms into the oral cavity, behind the posterior edge of the hard palate: a 8.5 mm 30° degree endoscope, a 5 mm EndoWrist® Maryland dissector and a 5 mm EndoWrist® Monopolar cautery, respectively positioned on the middle, right and left labial commissures. The endoscope looks upwards. Four phases can be described:

Mucosal phase: once the endoscope is placed beyond the hard palate, an inferior view of the cavum and the choanae is observed. The surgeon at the console performs an inverse U-shape flap of the posterior cavum mucosa, which corresponds to the mucosa covering anteriorly and inferiorly the sphenoidal rostrum (*see* Fig. 11). It is mandatory to see a

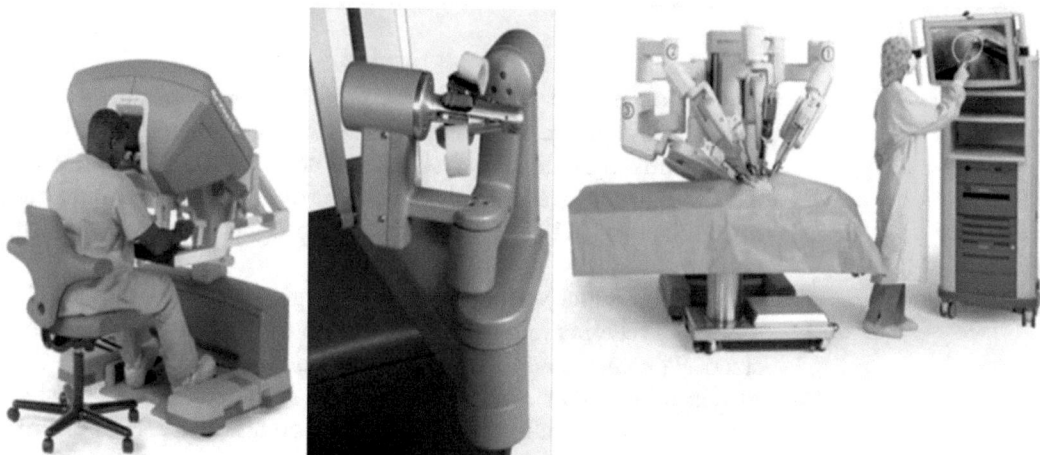

Fig. 8 The da Vinci system components, from left to right: the surgical console where the surgeon sits, the masters controllers for the operator, the patient cart with the 4 robotic arms and the vision cart (courtesy of Intuitive Surgical, Inc.)

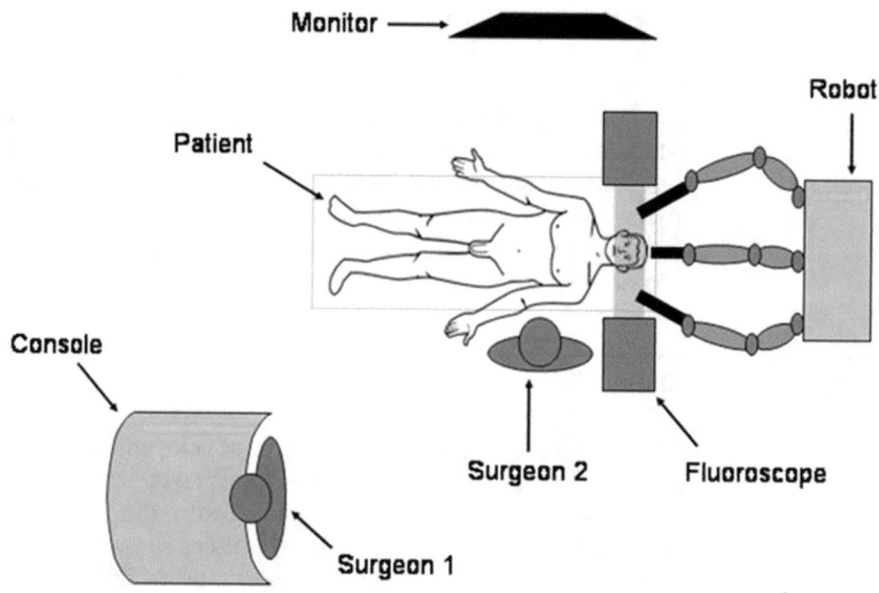

Fig. 9 Schematic view of the operative room. Surgeon 1 is the head and neck surgeon working at the console (SH); surgeon 2 is the neurosurgeon working at the bedside (DC) [7]

specific key point, which corresponded to the junction between the vomer and the sphenoid (*see* Fig. 11d). This point has to be the entering point to open the sphenoid bone safely.

Sphenoid phase: the surgeons' roles change and one robotic arm is removed. The endoscope remains at the midline and the second instrument is used to maintain the flap or perform additional suction. As the da Vinci system has no bony instruments,

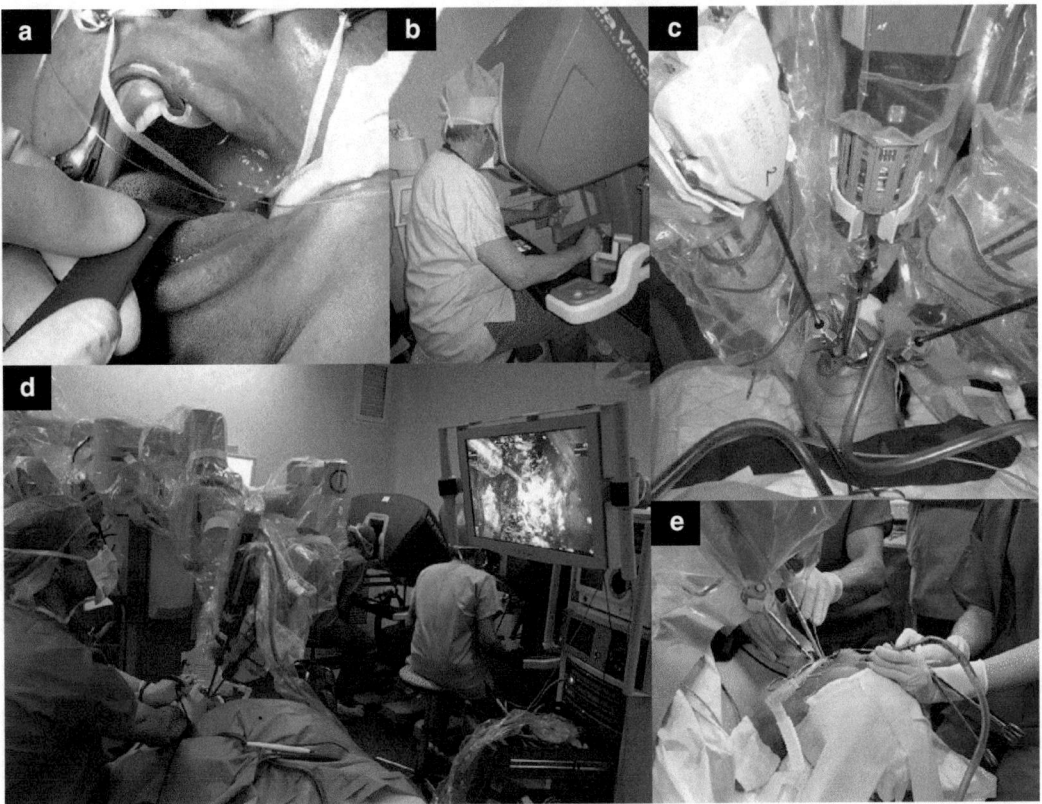

Fig. 10 Intraoperative photographs. (**a**) Mouth exposure with loops retracting the soft palate on each side of the uvula, which has been sutured (patient n°2). (**b**) The head and neck surgeon (SH) performs the mucosal dissection at the console of the robot. (**c**) Operative view during the mucosal time (patient n°3); the three arms are inserted into the mouth cavity and the tongue is retracted downwards with a suture. (**d**) and (**e**) Operative view during the sphenoid time (patient n°3 and 4); the neurosurgeon (DC) performs the drilling at the bedside with his two hands placed at the labial commissure. An additional suction can be placed in the nasal cavity

the opening of the sphenoid sinus is performed by the neurosurgeon at the bedside, watching his dissection on the 2D flat-panel screen. The drilling of the key point is performed with a *Midas* Rex® Legend *Stylus*®, using an angled handpiece. We use 3 and 5 mm diamond matchstick burs. The sphenoid sinus is then enlarged with Kerrison punch to maximise visualization of the sella turcica (*see* Fig. 12).

Sellar phase: the sellar floor is opened with the same drilling system. For the dural opening, we use a CO_2 flexible Laser (Luminis®) guided by the robotic instruments (*see* Fig. 12), rather than the Monopolar cautery of the da Vinci. Finally the tumour removal is performed with curettes by the neurosurgeon at the patient's side.

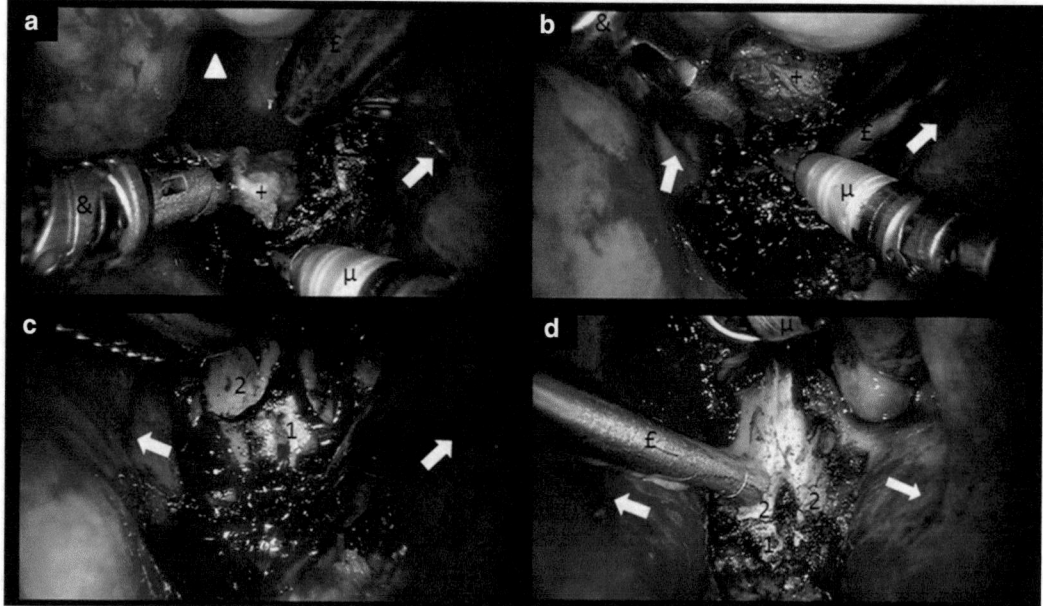

Fig. 11 Intraoperative view at the console during mucosal time. (**a**) and (**b**) The mucosal flap (+) is progressively dissected and retracted upwards using the Maryland dissector (&) and the Monopolar cautery (μ) (patient n°1). (**c**) visualization of the junction between the vomer, with its two alae (2), and the sphenoid bone (1) (patient n°1). (**d**) suction (£) showing the key point to enter the sphenoid sinus (patient n°2). White triangle: right choanae; white arrows: Eustachian tubes. [ref]

Closure phase: after tumor removal, oxidized regenerated cellulose is placed against the sellar aperture and the mucosal flap is reapplied, and sometimes glued.

Postoperative CT-scan were performed showing the inferior approach of the sella with TORS (*see* Fig. 13).

5   Tips and Tricks

The first major concern is selection of appropriate patients. Three factors seem important: the well-pneumatized sphenoid sinus, the mouth aperture quality and the projection line on the sella. Indeed it is mandatory to exclude patients with conchal sinus and patients suffering from trismus. With simple landmarks on an open mouth CT-scan, one can assess sellar accessibility at the projection line (see above).

During the surgical setup, if the exposition on the cavum is not large after soft palate retraction, one can put a supplementary stitch on the tongue to leave space for the robotic arms in the oral cavity (*see* Fig. 10c). This stitch will not cause any harm.

Once the three arms are correctly placed behind the soft palate, it is possible to perform infiltration of the mucosa to prevent from

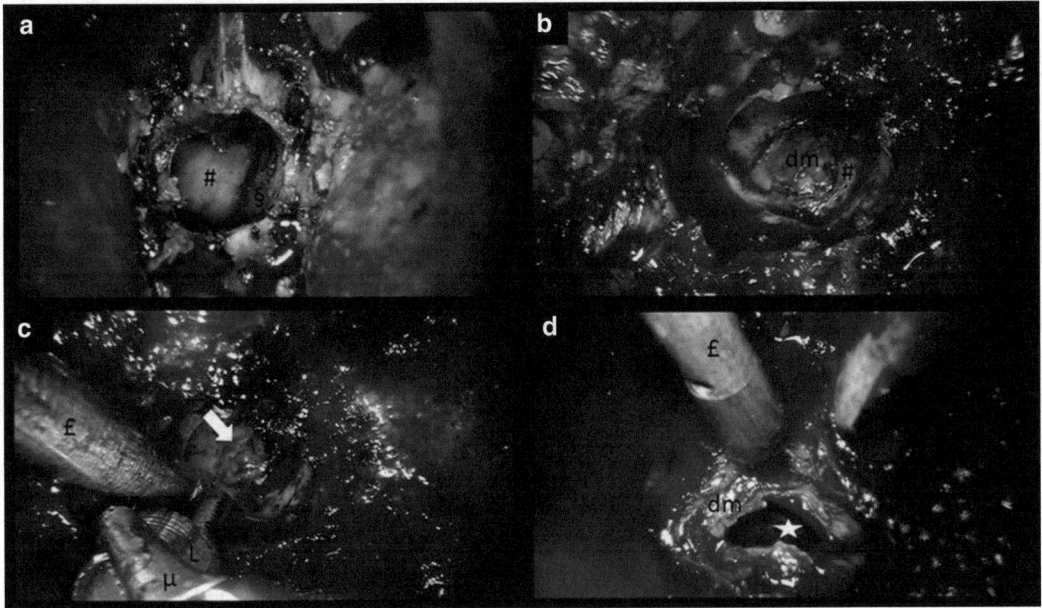

Fig. 12 Intraoperative view after drilling time. (**a**) general view on the sphenoid sinus and the sella (#) before opening of the intersinus sphenoid septum (§) (patient n°2). (**b**) and (**c**) sellar time including opening of the dura (dm) with the CO_2 laser fiber (L) handled by the Monopolar cautery (μ) (patient n°4). White arrow shows the X-shape dural aperture and the cyst evacuation. £: suction. (**d**) final view after tumour resection showing the sellar diaphragm (white star) (patient n°2)

bleeding. We use xylocaine and adrenaline to do so. We found it efficient but we note that it can also narrow the operative corridor.

Concerning the mucosal flap, we initially dissected a U-shape flap on cadaveric specimens. On patients we chose to perform an inverse U-shape flap in order to fill the sella in case of CSF leak. Thus we kept it in the right choana with a robotic instrument.

During the drilling, it is mandatory that the dissection remains in the midline, as with all transsphenoidal approaches. As we place the drill at the right labial commissure, the dissection is naturally directed towards the contralateral structures and thus to the left part of the sella. Unfortunately the da Vinci has no integrated neuronavigation system, hence we often looked backwards with the endoscope to check if the drilling was still at the midline and we especially studied before surgery the anatomy of the intersinus sphenoid septum.

6   Conclusion

We have created an innovative TORS to reach sellar tumors with promising results, especially on cystic lesions. We emphasize the minimal invasiveness of this approach and the new inferio-superior

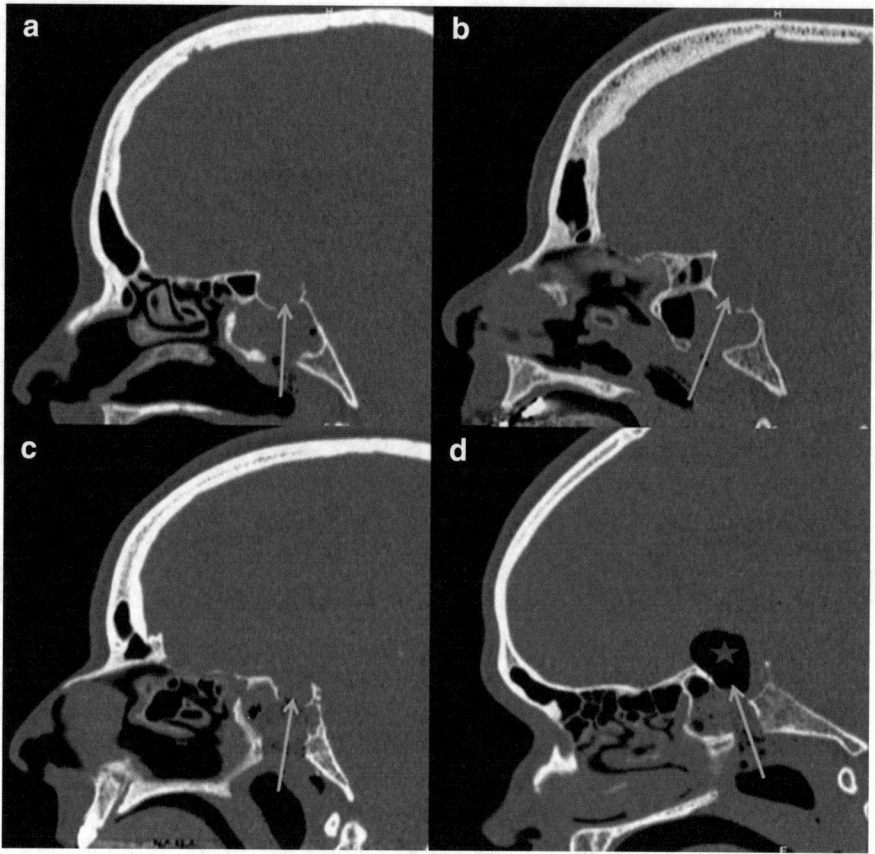

Fig. 13 Postoperative brain CT scans at day 1 showing the infero-superior approach of the sella with green arrows (**a**, **b**, **c** and **d** for patients n°1, 2, 3 and 4 respectively). Red star indicates postoperative pneumocephalus in the pituitary fossa for patient n°4. [ref]

direction to remove pituitary adenomas. The latter could potentially offer better access for large suprasellar extension of tumors, eliminating the need for transcranial surgery. However we must moderate our conclusions with several points. We cannot perform a complete robotic procedure as such, because the da Vinci system was designed for general surgery. Thus, we lack some dedicated neurosurgical instruments and the pituitary adenoma removal does need specific tools, like curettes. Moreover no bony instruments exist on the robot, except for a prototype, called "Endowrist rongeur", which has been tested on pigs for laminectomy [14]. A drilling system would need to be developed for the da Vinci if neurosurgeons hope to explore the field that we have just cleared further. At present, the da Vinci has no integrated neuronavigation system. But one can easily imagine that augmented-reality surgery [15, 16] could be applied to our skull base field. Indeed, contrary to other organs in the abdomen, the skull base constitutes a hard and still structure. It would be interesting to incorporate a projection of carotid arteries, cavernous sinus, etc. into the stereo viewer of the console. Finally, in order to overcome these challenges, we must look at different existing robotic systems (or invent it) that could allow a better TORS. The perfect teleoperated robot would have to offer a drilling system, dedicated microscopic instruments, integrated neuronavigation (or augmented-reality technique) and haptic feedback.

Acknowledgments

We thank Pr Lot for his general support regarding this project from the beginning to the last patients. We also thank Antoine Missistrano whose surgical assistance was precious and Wendy Gold for English corrections.

References

1. Hans S, Delas B, Gorphe P et al (2012) Transoral robotic surgery in head and neck cancer. Eur Ann Otorhinolaryngol Head Neck Dis 129:32–37

2. Hans S, Jouffroy T, Veivers D et al (2013) Transoral robotic-assisted free flap reconstruction after radiation therapy in hypopharyngeal carcinoma: report of two cases. Eur Arch Otorhinolaryngol 270:2359–2364

3. Lee JY, O'Malley BW Jr, Newman JG et al (2010) Transoral robotic surgery of the skull base: a cadaver and feasibility study. ORL J Otorhinolaryngol Relat Spec 72:181–187

4. Yang MS, Yoon TH, Yoon DH et al (2011) Robot-assisted transoral odontoidectomy: experiment in new minimally invasive technology, a cadaveric study. J Korean Neurosurg Soc 49:248–251

5. Marcus HJ, Hughes-Hallett A, Cundy TP et al (2015) Da Vinci robot-assisted keyhole neurosurgery: a cadaver study on feasibility and safety. Neurosurg Rev 38:367–371

6. Crockard HA (1985) The transoral approach to the base of the brain and upper cervical cord. Ann R Coll Surg Engl 67:321–325

7. Chauvet D, Missistrano A, Hivelin M et al (2014) Transoral robotic-assisted skull base surgery to approach the sella turcica: cadaveric study. Neurosurg Rev 37:609–617

8. Amelot A, Trunet S, Degos V et al (2015) Anatomical features of skull base and oral cavity: a pilot study to determine the accessibility of the sella by transoral robotic-assisted surgery. Neurosurg Rev 38:723–730

9. Chauvet D, Hans S, Missistrano A et al (2017) Transoral robotic surgery for sellar tumors: first clinical study. J Neurosurg 127:941–948

10. Liu JK, Das K, Weiss MH et al (2001) The history and evolution of transsphenoidal surgery. J Neurosurg 95:1083–1096

11. Carrau RL, Prevedello DM, de Lara D et al (2013) Combined transoral robotic surgery and endoscopic endonasal approach for the resection of extensive malignancies of the skull base. Head Neck 35:E351–E358

12. Haidegger T, Sándor J, Benyó Z (2011) Surgery in space: the future of robotic telesurgery. Surg Endosc 25:681–690

13. Vasilescu D, Paun S (2012) Surgical treatment of parietal defects with "da Vinci" surgical robot. J Med Life 5:232–238

14. Ponnusamy K, Chewning S, Mohr C (2009) Robotic approaches to the posterior spine. Spine 34:2104–2109

15. Liu WP, Richmon JD, Sorger JM et al (2015) Augmented reality and cone beam CT guidance for transoral robotic surgery. J Robot Surg 9:223–233

16. Porpiglia F, Bertolo R, Amparore D et al (2018) Augmented reality during robot-assisted radical prostatectomy: expert robotic surgeons' on-the-spot insights after live surgery. Minerva Urol Nefrol 70:226–229

Chapter 12

Robot-Assisted Pedicle Screw Placement

Florian Roser and Nader M. Hebela

Abstract

With new technological advancements in spine surgery, come accompanying expectations regarding the association between the incremental improvements of the surgical technique and its ability to provide superior patient outcomes and more efficient surgical workflows at lower cost. In many surgical fields, robots are still being tested on an experimental level. In others, such as urological surgery and gastrointestinal surgery, robots are a routine part of subspecialty practice. Spinal surgeons have begun to adopt the available robotic systems into their daily workflow. Yet, despite the majority of clinical results showing equally high or even higher accuracy for robot-assisted pedicle screw instrumentation and reduced radiation exposure, the evidence for long-term clinical outcome is still scarce. The chapter provides an overview of the existing robotic platforms and the workflow in spine surgery, with an emphasis on minimal invasive fusion procedures, where integration of robotic systems might be especially beneficial. Future technological improvements in these robotic systems specific to spine surgery, the integration of intraoperative imaging modalities, and the awareness of a significant learning curve will provide a cost-effective surgical tool that facilitates better clinical results with decreased surgical time.

Key words Spinal robotics, Spinal instrumentation, Minimal-invasive techniques, Percutaneous

1 Introduction

During conventional open spine surgery, consisting mainly of decompression and stabilization, the target structures and trajectories are identified by exposure of the surface anatomy in combination with 2D fluoroscopy. High-grade degenerative disease, complex spinal deformity, revision surgery, and the increasing number of minimal invasive techniques pose a challenge for the anatomical orientation, even for the experienced surgeon, using traditional visualization of surface landmarks and 2D fluoroscopic techniques. In this respect, robotics for spinal surgery holds great promise. Robotic-assisted spine surgery has the potential to increase the accuracy and precision of the surgeon, particularly in

Electronic supplementary material: The online version of this chapter (https://doi.org/10.1007/978-1-0716-0993-4_12) contains supplementary material, which is available to authorized users.

Hani J. Marcus and Christopher J. Payne (eds.), *Neurosurgical Robotics*, Neuromethods, vol. 162, https://doi.org/10.1007/978-1-0716-0993-4_12, © Springer Science+Business Media, LLC, part of Springer Nature 2021

the context of variable anatomy due to loss of native landmarks in revision surgery and abnormal anatomy in deformity and high-grade degenerative changes. As the use of robotics in surgery is becoming more popular and widespread the utility is becoming more evident in spinal surgery.

Several 3D computer-assisted navigation options are currently available for use in the field of spine surgery: (1) Airo Mobile intraoperative computer tomography (CT)-based Spinal Navigation (Brainlab, Feldkirchen, Germany); (2) Stryker Spinal Navigation with SpineMask Tracker; SpineMap Software (Stryker, Kalamazoo, Michigan); (3) Stealth Station Spine Surgery Imaging and Surgical Navigation with O-arm (Medtronic, Minneapolis, Minnesota); (4) Ziehm Vision FD Vario 3-D with NaviPort integration (Ziehm Imaging, Orlando, Florida).

In many ways, spine surgery is ideally suited for the integration of robot-assisted surgical procedures. Although navigation assists the surgeon in display of trajectories in a complex 3D environment, the surgical task, e.g. cannulating the pedicle, has to be carried out "free-hand". This manipulation has to be performed in close proximity to critical structures that are more frequently accessed through minimally invasive approaches. A robotic interface actually represents the missing link between pre-operative imaging, trajectory planning and surgical execution thereby significantly improving repetitive tasks with precision and reproducible results [1, 2].

2 Summary of Evidence

The high volume of reported data relevant to the topic of pedicle screw accuracy and safety of implantation is ideal for meta-analysis comparison. A meta-analysis published by Kosmopoulos and Schizas, covering 37,337 pedicle screw implants in total, determined in a subgroup of in vivo implants a median accuracy of 95.1% and 90.3% with and without the assistance of navigation, respectively (15,358 screws) [3]. Verma et al. conducted the first of several meta-analyses on the topic of robotic-guided surgery including 23 studies, evaluating 5992 pedicle screws and found a significantly higher rate of accuracy utilizing computer-assisted navigation. However, though the rate of neurological injury favoured navigation, the group failed to demonstrate statistical significance ($P = 0.07$) [4]. On the other hand, no statistical significant difference could be found for complication rates and the study was unable to make conclusion regarding functional outcome. The findings of a later meta-analysis performed by Shin et al. in 2012 confirmed this trend, including over 7000 pooled pedicle screws. Additionally, Shin et al. performed sub-analyses of cervical, thoracic, and lumbar pedicle screws and reported significantly more accurate screw placement for all 3 spinal segments utilizing navigation [5]. Gelalis et al. reviewed 26 studies and 6617 pedicle screws inserted free-hand, with fluoroscopic guidance or with computer-assisted surgery [6]. While they found no significance between the fluoroscopic and navigation-assisted methods, both exhibited statistically superior accuracy as compared to free-hand technique. However, all authors failed to demonstrate a significantly lower rate of screw revision or total reoperation rates and any difference in neurological injury. In recent years, the implementation of intraoperative imaging has further refined the possibilities and helped to overcome some drawbacks of image-guided navigation for spinal instrumentation. Intraoperative 3D-fluoroscopy and computed tomography (CT) increases accuracy of navigation system and facilitates the registration process [7]. While the benefits of intraoperative image-guided surgery are evident, there is further demand for alternative active assistance in the surgical workflow of spinal procedures.

Currently, SpineAssist and Renaissance Surgical Guidance Robot (Mazor Robotics, Ltd., Casesarea, Israel), ROSA (Zimmer Biomet, Warsaw, IN, USA), and the recently added ExcelsiusGPS (Globus Medical, Inc., Audubon, PA, USA) are the only robotic systems that are FDA-approved for pedicle screw trajectory guidance. With a marketed retail base price of $850,000 and annual disposable expenses estimated to be nearing $2000, concerns for cost-effectiveness are evident [8]. The ROSA Robot, originally designed for cranial stereotactic procedures [9], comes with a flat-panel CT device for its spinal application. On 38 implanted screws in a cadaver model it demonstrated a 97.4% accuracy [10]. This preliminary study was followed by patient experience with comparable excellent results if the robot was coupled to flat-panel computed tomography guidance [11]. The newest member of approved devices is the ExcelsiusGPS robot with preliminary clinical data available [12].

Since 2005, numerous cadaveric and clinical studies on the SpineAssist system have been evaluating the technical feasibility, radiation exposure and the accuracy of pedicle screw instrumentation at the thoracic, lumbar and sacral spine [13–21]. A recent meta-analysis by Marcus et al. evaluated all available studies, including two randomized, two cohort studies and one cadaver study [13, 15–17, 22], comparing pedicle screw placement with robot-assistance vs. fluoroscopy guidance [20]. Among the total of 1308 pedicle screws, 729 were instrumented robot-assisted and 579 under fluoroscopy-guidance with a satisfactory position of 94.1% and 92.7%, respectively. A retrospective, multi-centre evaluation of 3271 SpineAssist-supported spinal implantations between 2005 and 2009, involving a large percentage of paediatric scoliosis patients, reported a clinically acceptable placement rate of 98% assessed by intraoperative fluoroscopy [23]. Postoperative CT-based evaluation of a subset of patients included 646 pedicle screws and demonstrated an accuracy of 98.3% according to Gertz-bein and Robbins A/B criteria [24], with a mean axial and sagittal plane deviation of 1.2 ± 1.49 mm and 1.1 ± 1.15 mm, respectively. Only two screws deviated >4 mm from the pedicle wall, but without irreversible neurological deficits reported. Pechlivanis et al. in their prospective study, reported successful use of the device in 31 cases, with 98.5% of screws demonstrating axial and longitudinal accuracy, i.e. deviations less than 2 mm [25]. Another case series evaluated a total of 960 pedicle screws in 102 patients, primarily presenting spinal deformities and/or revision surgeries. Accurate positioning was achieved in 98.9% of the screws [26]. Eleven screws were considered misplaced, of which ten screws were manually corrected during surgery. One patient required implant removal within 3 days of surgery as a result of radiculopathy. Screw misplacement was presumably caused by tool "skiving", i.e. the tip of the drill holder or guiding tool skids off the intended entry point leading to an aberrant trajectory. Similar aspects for inaccurate screw placement have been pointed out by Ringel et al. [15]. In contrast to all other comparative data this randomized, prospective study assessed screw accuracy in favour of free-hand fluoroscopy-guided over robot-assisted instrumentation with a satisfactory placement rate (Gertzbein and Robbins A/B on postoperative CT) of 93% and 85%, respectively. They concluded that besides displacement of the entry point by soft tissue pressure, an unstable attachment of the robot to the spine may have contributed to the assessed inaccuracies.

During conventional fluoroscopy-guided pedicle instrumentation the patient and the surgical team are exposed to a significant amount of radiation. Several clinical studies have shown that robot-assisted instrumentation decreases the occupational risk for surgeons and operation room staff members by significantly reducing intraoperative radiation doses, particularly in minimally invasive surgery (MIS) procedures [17, 22, 27]. A retrospective cohort analysis comparing conventional open to robotic-guided open/percutaneous pedicle screw insertions measured up to 70% less radiation exposure in the robot-guided procedures, 64% of which were performed using percutaneous approaches, when compared to 57 free-hand surgeries, all executed using an open approach [22]. Similarly, in a prospective, randomized comparative trial of freehand, navigation-guided and robotic-guided spinal procedures, a twofold reduction in radiation time and dose when integrating the robotic system to insert pedicle screws in a minimally invasive approach was reported, as compared to freehand surgeries in a conventional open approach [13]. In another single-centre study of conventional versus robotic-assisted percutaneous spine fusion surgeries, a median 40% lower radiation exposure was measured for the robot cohort [28]. Upon implementation of SpineAssist in a controlled, cadaveric implantation trial, overall radiation exposure among 87% of the surgeons was below 1mrem, versus a mean 13 6mrem exposure when using traditional surgical approaches [17]. When calculating the average radiation per screw, a 98.2% reduction in ionization exposure was observed in robot-guided procedures, with an average 0.2mrem per screw, versus a mean 10.1mrem per screw in the control group. Fluoroscopy time was reduced by a similar degree, from a mean 33.0 s per screw in the control group and 0.7 s per screw.

The present experience confirms a high level of accuracy for robot-assisted pedicle instrumentation for both the conventional open and percutaneous approach. In more traditional open cases in which the anatomy is clearly exposed, the advantage is less significant and the extra input might not be justified. Whereas in high-grade degenerative disease, spondylolysis, revision surgery and in complex deformity cases, visualization of surface anatomy alone may not be sufficient, and robotic guidance can be helpful. Robotic systems can guide the surgeon to the precise location without needing to visualize the anatomy. These robotic systems should have clear advantages in percutaneous or MIS procedures [22]. A multi-centre evaluation of robot-guided cases by DeVito, in which a significant subgroup had complex deformities, showed that 49% of the screws in their study were placed in a percutaneous approach [23]. MIS procedures in spine surgery have gained more acceptance and are applied more frequently as they are associated with less postoperative pain, lower infection rates, less blood loss and less paraspinal muscle trauma, reduced recovery period and tissue scarring [29–31]. On the other hand, MIS procedures are usually associated with increased operating time and have been reported to expose patients and surgeons to higher radiation doses [17, 22, 27, 32, 33]. For instance, interbody fusion techniques represent, besides pedicle instrumentation, an integral part of treatment concepts for segmental spinal instability as a frequent aspect of progressive lumbar degeneration, in isthmic spondylolisthesis, post-laminectomy situations and pseudo-arthrosis [34, 35]. TLIF (transforaminal lumbar interbody fusion) is particularly suited for minimal invasive or percutaneous approaches due to its posterior, unilateral access to the spine [35]. Numerous studies comparing, MIS TLIF versus open TLIF, have been performed in recent years and results are fairly homogeneous [32, 33, 36–40]; the clinical outcome reported is generally favorable with reduced postoperative pain, reduced intraoperative blood loss and shortened hospital stay. A meta-analysis by Karikari et al. quantified the blood loss from 150 to 456 mL for MIS TLIF, and 366.8 to 1147 mL for open TLIF. The duration of postoperative hospital stay ranged in the MIS TLIF from 3.0 to 10.6 days and from 4.2 to 14.6 days for open TLIF [41]. The clinical outcomes range from favourable to excellent (assessed by VAS and ODI), especially during the immediate postoperative course [38], although long-term outcome is not significantly better for MIS than for open TLIF [32, 37, 39]. Studies from Peng et al. and Wang et al. found both operating time and radiation exposure to be increased in MIS TLIF procedures [32, 33]. Concerns whether equally solid bony fusion is feasible in MIS TLIF, in which preparation of the disc space through narrow corridors is more challenging, have been questioned by numerous studies [36, 37]. Recently Wu et al. reported in a meta-analysis a fusion rate of 94.8% and 90.9% for MIS versus open TLIF [42].

3 Description of Robot

The Renaissance Surgical Guidance Robot and its predecessor SpineAssist, both available for routine clinical use in spinal procedures, are examples of supervisory-control systems. The robot is a cylindrical (50 × 90 mm) device equipped with an end-effector with six degrees of freedom. The bone-mounting platform interfaces the robotic device with the patient's skeleton and the robotic arm is positioned according to pre-planned, image-guided trajectories while the surgeon directs the instruments (e.g. drill guide, cannulas) along the pre-defined path. Applying this concept, the device is not only applicable for spinal instrumentation but also for collecting biopsies, tumour excisions, cement augmentations and extraforaminal disc prolapses in distorted anatomic spaces [13, 14, 22, 23, 26].

Since the commercial introduction of the device numerous studies have outlined the accuracy and precision. In our own series [13], postoperative CT-scans were assessed for screw placement accuracy. 96% of the overall 120 screws met Gertzbein and Robbins criteria A/B. Misplaced screws (grades C/D, 3 screws) were located at the lateral pedicle wall, but were not associated with neurological deficits as the deviation was to the lateral superior aspect of the pedicle. The overall mean operation time was 165 min and mean blood loss was assessed to be 162 mL, which are on the lower range of the previously reported MIS procedures. Under robot-guidance a mean of 17 min was needed for k-wire placement and further 15 min for final screw insertion. Radiation time per screw was assessed with a mean of 4.1 s and radiation dose was 3.4 mGy per screw. Observed complications included one patient with postoperative epidural hematoma that needed evacuation and one patient with an L5-radiculopathy postoperatively that was not related to screw misplacement. No infections were observed.

A similar prospective randomized study on 78 patients with robotic assisted and freehand technique were published in 2017 [43]. Robotic-assisted pedicle screw placement was associated with similar intrapedicular accuracy to the freehand technique, fewer proximal facet joint violations, higher screw positioning accuracy for the proximal facet joint, and less hazardous orientations.

4   Description of Technique

The surgical workflow involves acquiring a CT of the patient's spine for the preoperative planning stage. In the operating room, the robotic platform is attached to the patient's spine involving pins to the spinous process and to the iliac crest or to the OR table. Two fluoroscopic images (anteroposterior and 60° oblique to the lateral plane) of the spine and the robotic platform (marked by a 3D fiducial array) are used for registration and matching to the preoperative CT (Fig. 1). Following 3D synchronization, the robot is attached to the platform and the robotic arm is dispatched to the calculated trajectories (Fig. 2). After stab skin incisions (1–1.5 cm) and robot-assisted Kirschner-wire insertion, the four pedicle screws are placed percutaneously followed by removal of the robotic platform. In the majority of cases the rod construct is inserted and temporarily tightened (Video S1). This is followed by either a small midline incision (about 3 cm) or a lateral incision over the already existing percutaneous screw insertion sides. The next step is the subperiosteal placement of the retractor system to gain access to the ipsilateral lamina and facet joint. Microsurgical facetectomy is performed using high-speed burr and rongeurs. Bone may be harvested for graft material. If indicated, bilateral decompression of the dural sac and the exiting nerve root is feasible through the unilateral approach. Eventually, lateral exposure of the annulus allows for meticulous preparation of the disc space and

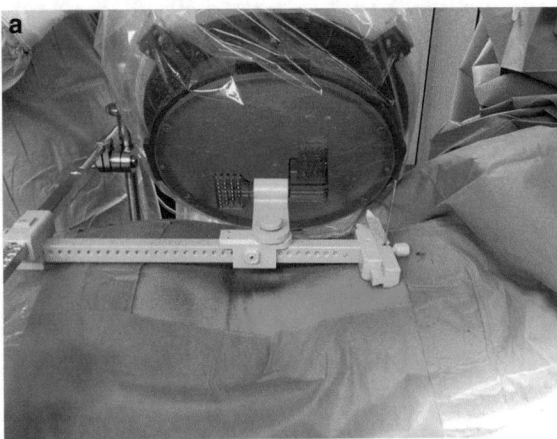
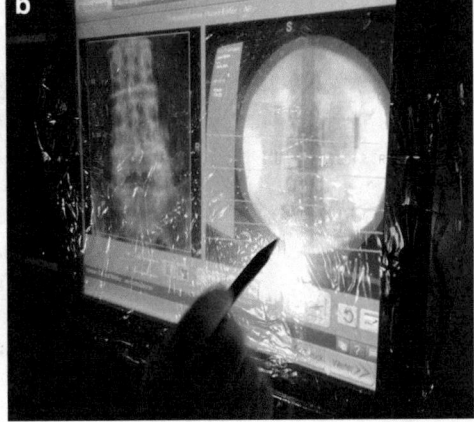

Fig. 1 (a, b) Two fluoroscopic images (anteroposterior and 60° oblique to the lateral plane) of the spine and the robotic platform (marked by a 3D fiducial array) are used for registration and matching to the preoperative CT

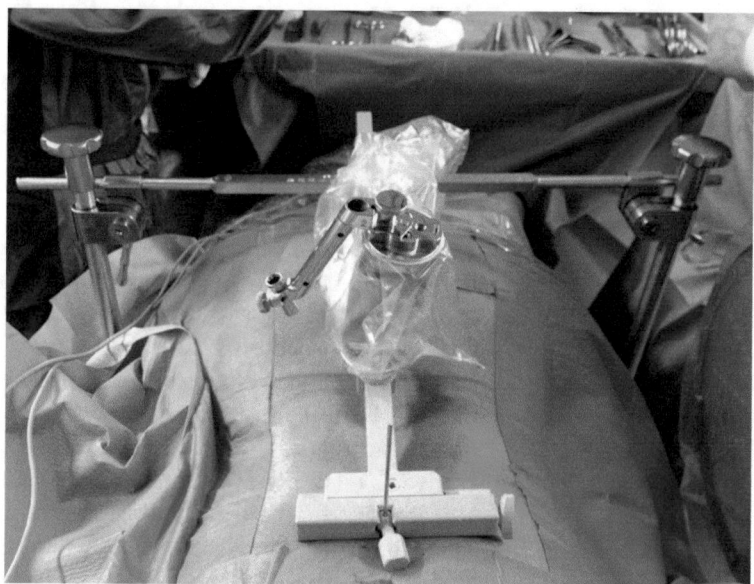

Fig. 2 The robot is attached to the platform and the robotic arm is dispatched to the calculated trajectories

implantation of a TLIF cage is performed under fluoroscopic control. This is followed by compression of the screw-rod system and final tightening of the screw-caps.

5   Tips and Tricks

Implement navigation into your everyday easy cases, initially. For example, use navigation for cases where you may not even use pedicle screws, such as basic decompressions. While it will initially feel like a triumph of technology over reason, establishing a workflow with simpler cases will make the more complicated ones easier to perform.

When using intraoperative navigation, make sure you follow a basic line-of-site rule: the reference frame should always be between the surgeon and the camera or optical device. For example, when performing lumbar instrumentation, place the optical scanner at the foot of the bed and the reference frame in the iliac crest, while the instrumentation is in the lumbar spine. Similarly, when performing cervical instrumentation, place the camera at the head of the bed and attach the reference frame to the C2 or C3 spinous process, as you instrument the subaxial cervical spine.

Place pedicle screws furthest away from the optical scanner first. This will ensure that the placement of additional screws is not made more difficult by the existing construct. For example, if instrumenting the lumbar spine, the order of screw placement should be L5 then L4 then L3. In the cervical spine, placement should be C3 then C4 then C5 pedicle screws. Use a "no-hands" technique without additional pressure, especially in the subaxial cervical spine where the motion between cervical levels in greater than in the lumbar spine. This is because the only truly accurate level is the one on which the reference frame is placed.

When you are unsure about the accuracy of the navigation, place a probe on a portion of the spine that is visible to confirm accuracy and to ensure what you are seeing on the navigation screen is what you see on the spine.

Using shared-controlled robotic systems, percutaneous skiving becomes an issue, with the sleeve deviating from the trajectory. Entry zones on bony surfaces should be planned perpendicular. However percutaneous planning software does not necessarily account for this situation nor can it be completely avoided during execution of the drilling as there is no visualization. Advanced robotic tools contain drill sleeves which are crowned to sink into the bone, therefore avoiding deviation from trajectory during drilling.

In general, robotic system tend to create trajectories which would not be the choice of an open-trained manual surgeon, therefore it becomes a confidence issue executing them. Try to plan "real life" trajectories during the blueprint of the robotic plan to ensure that if the case is converted to standard open instrumentation, the exposure or trajectory can still be used.

All robotic or navigated procedures harbor the risk of execution failure due to multiple sequential steps which could multiply errors leading to substandard precision. Despite using latest technologies, try to keep your workflow simple, reassure yourself using common sense, double-check each step and do not rely on a system you do not understand entirely.

6   Conclusions

Despite reported benefit, the use of robotics in surgery is still at an early stage. Robotics are not the standard of care in most fields. This is due to a number of reasons—cost/benefit trade-offs, familiarity, and ease of integration being several primary factors. This is to be expected when adopting new tools, especially ones of a deeply technical nature like robotics. New techniques require new knowledge and a slightly different skill set, it may necessitate new approaches and therefore harbours increased opportunities for equipment failures. As these new techniques also disrupt existing workflows, they alter communication and teamwork patterns and therefore demand new ways of thinking. Workflow disruptions, albeit vaguely defined, can be lead to altered visual access, non-verbal cues, less eye contact. Shared mental models are necessary to anticipate needs and can improve response time five-fold [44]. One study found that new team members (and new technology can be considered as a new team member) are responsible for 20–30% of interruptions and miscommunications [45].

Image-guided navigation in spinal surgery has demonstrated its feasibility and qualities for spinal instrumentation in numerous studies, but lacks sufficiently powered data to prove the benefit on clinical outcome. Correspondingly, the available, mainly observational, data establish a superior accuracy for robot-assisted pedicle instrumentation in the thoracic, lumbar and sacral spine when compared with conventional surgery in addition to reduced radiation exposure. The prospective RCTs available to date, however, show inconsistent results [13, 15, 43, 46]. Further prospectively and randomized controlled data on accuracy and especially patient outcome are still needed for the evaluation of robotic guidance systems.

A key goal of recent robotic systems' development has been to improve on the present state of the art to achieve continuous, real-time, load-controlled or hybrid-controlled (as opposed to position-controlled) testing and to apply it in various scenarios approximating physiological ranges of motion (ROMs). An important advantage of robot-based spinal testing systems is multiaxial motion, which allows for reorienting force vectors in real time to better simulate in vivo conditions. Two publications from a Memphis based group highlighted the development and application of a novel cartesian-based real-time load-controlled robotic testing system [47, 48]. Their validating experimental application exemplifies the advantages of robotic testing over non-robotic platforms; namely to apply a dynamically oriented force. Their system is capable of applying a real-time dynamically reorienting force that follows the midline of the intervertebral disc as the joint of the specimen moves with rotation and deformation, which allows rigorous testing throughout the physiological ROM [49].

The financial investment in a surgical robot or other computerized surgical systems is substantial. While navigation systems were introduced over two decades ago, and have been proven to produce superior results, only 11% of spine surgeons responding to a survey reported using such technology in their surgical protocols [50]. Cost efficiency is a critical factor for the appraisal and adoption of new technologies in the clinical setting, and expectations will further increase. If the clinical value of computer-assisted spinal surgery, for example in MIS TLIF surgery, which is promising but has to be further validated, consists in significantly reduced hospital stay, reduced convalescence and postoperative pain and thus improved patient outcome, general acceptance and financial compensation will escalate. It is imperative for spine surgeons to understand the economical dynamics of this new technology before integrating it into their institution or hospital's surgical system.

Though many sources have commented on the technology being cost-effective, the spine surgery literature on the use of robotics fails to back this claim, for now. Some studies have validated accuracy and effectiveness while simultaneously seeing a reduction in fluoroscopy time, OR time, and revision rates. These findings make a case for cost savings, but real time data validating these claims over conventional means of instrumentation (both traditional open techniques as well as MIS) is lacking. Also, there is no data on the use of robots in spine surgery regarding the effects on the length-of-stay, as a decrease in length-of-stay will have a direct effect the cost burden and expenditure. A decrease in post-operative pain has been established with MIS procedures, but there has not been a retrospective study or a prospective trial that has investigated whether robotic technology has an effect on the post- operative stay in the hospital.

Robotic-assisted spinal surgery, though proven to be safe and efficacious for pedicle screw instrumentation, has an even larger cost burden to overcome prior to widespread adoption. Results of pedicle screw accuracy utilizing these robotic arms are at best equal to those reported for computer-assisted surgery alone and currently add little to the effort to improve safety and effectiveness of spinal surgery [51, 52]. For future applications of robotic systems, it will be critical to provide non-fluoroscopic intra-operative imaging such as a miniaturized MRI that can be located within the OR. Reduced size of the tele-robotic console or augmented reality could improve real time situation awareness. Force and/or torque sensors by adding sensor technologies such as force-based tools on autonomous robots could detect plunging/breach to minimize injury and would become *smart* and lead the way to the next level of robot-assisted spine surgery. Shared-controlled in future applications means that tactile feedback and *feel* is largely maintained. As important as the focus is on engineering and technology of new systems, the success depends on the ease of use, the acceptance by surgeons and OR staff and the increased efficiency that it can offer to the surgical team. Studies looking at low-profile, flexible robotic arms attached to the ceiling, incorporating image-guided navigation, real-time electrophysiological data to clear the operating field are examples on how design can improve surgeons experience. The more involved the neurosurgeon remains in the robotic surgery workflow the more likely these technologies will gain acceptance. The biggest challenges in this field however remains error detection. So far, interfaces are fairly intuitive and new robotic concepts contain innumerable steps to execute the tasks. Knowing when to trust or not trust the system and understanding how to employ redundancies and recover from failure is paramount to gain acceptance in the spinal surgical community [51, 52].

7   Electronic Supplementary Material

Video S1 Illustrative case (MP4 225,835 kb).

References

1. Kelly PJ (2002) Neurosurgical robotics. Clin Neurosurg 49:136–158
2. Louw DF, Fielding T, McBeth PB, Gregoris D, Newhook P, Sutherland GR (2004) Surgical robotics: a review and neurosurgical prototype development. Neurosurgery 54:525–536; discussion 536–7
3. Kosmopoulos V, Schizas C (2007) Pedicle screw placement accuracy: a meta-analysis. Spine (Phila Pa 1976) 32:E111–E120
4. Verma R, Krishan S, Haendlmayer K, Mohsen A (2010) Functional outcome of computer-assisted spinal pedicle screw placement: a systematic review and meta-analysis of 23 studies including 5,992 pedicle screws. Eur Spine J 19:370–375
5. Shin BJ, James AR, Njoku IU, Hartl R (2012) Pedicle screw navigation: a systematic review and meta-analysis of perforation risk for computer-navigated versus freehand insertion. J Neurosurg Spine 17:113–122
6. Gelalis ID, Pachos NK, Pakos EE et al (2012) Accuracy of pedicle screw placement: a systematic review of prospective in vivo studies comparing free hand, fluoroscopy guidance and navigation techniques. Eur Spine J 21:247–255
7. Scheufler K-M, Franke J, Eckardt A, Dohmen H (2011) Accuracy of image-guided pedicle screw placement using intraoperative computed tomography-based navigation with automated referencing. Part II: thoracolumbar spine. Neurosurgery 69:1307–1316
8. Hu X, Lieberman IH (2014) What is the learning curve for robotic-assisted pedicle screw placement in spine surgery? Clin Orthop Relat Res 472(6):1839–1844
9. Lefranc M, Capel C, Pruvot-Occean AS, Fichten A, Desenclos C, Toussaint P, Le Gars D, Peltier J (2015 Feb) Frameless robotic stereotactic biopsies: a consecutive series of 100 cases. J Neurosurg 122(2):342–352
10. Lefranc M, Peltier J (2015) Accuracy of thoracolumbar transpedicular and vertebral body percutaneous screw placement: coupling the Rosa® spine robot with intraoperative flat-panel CT guidance—a cadaver study. J Robot Surg 9(4):331–338
11. Chenin L, Capel C, Fichten A, Peltier J, Lefranc M (2017) Evaluation of screw placement accuracy in circumferential lumbar arthrodesis using robotic assistance and intraoperative flat-panel computed tomography. World Neurosurg 105:86–94
12. Zygourakis CC, Ahmed AK, Kalb S, Zhu AM, Bydon A, Crawford NR, Theodore N (2018) Technique: open lumbar decompression and fusion with the Excelsius GPS robot. Neurosurg Focus 45(VideoSuppl1):V6
13. Roser F, Tatagiba M, Maier G (2013) Spinal robotics: current applications and future perspectives. Neurosurgery 72(Suppl 1):12–18
14. Togawa D, Kayanja MM, Reinhardt MK, Shoham M, Balter A, Friedlander A et al (2007) Bone-mounted miniature robotic guidance for pedicle screw and translaminar facet screw placement: part 2—Evaluation of system accuracy. Neurosurgery 60: ONS129–ONS139; discussion ONS139
15. Ringel F, Stüer C, Reinke A, Preuss A, Behr M, Auer F et al (2012) Accuracy of robot-assisted placement of lumbar and sacral pedicle screws: a prospective randomized comparison to conventional freehand screw implantation. Spine (Phila Pa 1976) 37:E496–E501
16. Schizas C, Thein E, Kwiatkowski B, Kulik G (2012) Pedicle screw insertion: robotic assistance versus conventional C-arm fluoroscopy. Acta Orthop Belg 78:240–245
17. Lieberman IH, Hardenbrook MA, Wang JC, Guyer RD (2012) Assessment of pedicle screw placement accuracy, procedure time, and radiation exposure using a miniature robotic guidance system. J Spinal Disord Tech 25:241–248
18. Sukovich W, Brink-Danan S, Hardenbrook M (2006) Miniature robotic guidance for pedicle screw placement in posterior spinal fusion: early clinical experience with the SpineAssist. Int J Med Robot 2:114–122
19. Barzilay Y, Liebergall M, Fridlander A, Knoller N (2006) Miniature robotic guidance for spine surgery—introduction of a novel system and analysis of challenges encountered during the clinical development phase at two spine centres. Int J Med Robot 2:146–153

20. Lieberman IH, Togawa D, Kayanja MM, Reinhardt MK, Friedlander A, Knoller N et al (2006) Bone-mounted miniature robotic guidance for pedicle screw and translaminar facet screw placement: part I—technical development and a test case result. Neurosurgery 59:641–650; discussion 641–50

21. Shoham M, Lieberman IH, Benzel EC, Togawa D, Zehavi E, Zilberstein B et al (2007) Robotic assisted spinal surgery—from concept to clinical practice. Comput Aided Surg 12:105–115

22. Kantelhardt SR, Martinez R, Baerwinkel S, Burger R, Giese A, Rohde V (2011) Perioperative course and accuracy of screw positioning in conventional, open robotic-guided and percutaneous robotic-guided, pedicle screw placement. Eur Spine J 20:860–868

23. Devito DP, Kaplan L, Dietl R, Pfeiffer M, Horne D, Silberstein B et al (2010) Clinical acceptance ad accuracy assessment of spinal implants guided with SpineAssist surgical robot: retrospective study. Spine (Phila Pa 1976) 35:2109–2115

24. Gertzbein SD, Robbins SE (1990) Accuracy of pedicular screw placement in vivo. Spine (Phila Pa 1976) 15:11–14

25. Pechlivanis I, Kiriyanthan G, Engelhardt M, Scholz M, Lücke S, Harders A et al (2009) Percutaneous placement of pedicle screws in the lumbar spine using a bone mounted miniature robotic system: first experiences and accuracy of screw placement. Spine (Phila Pa 1976) 34:392–398

26. Hu X, Ohnmeiss DD, Lieberman IH (2013) Robotic-assisted pedicle screw placement: lessons learned from the first 102 patients. Eur Spine J 22:661–666

27. Rampersaud YR, Foley KT, Shen AC, Williams S, Solomito M (2000) Radiation exposure to the spine surgeon during fluoroscopically assisted pedicle screw insertion. Spine 25:2637–2645

28. Schoenmayr R, Kim I-S (2011) Why do I use and recommend the use of navigation? ArgoSpine News J 22:132–135

29. O'Toole JE, Eichholz KM, Fessler RG (2009) Surgical site infection rates after minimally invasive spinal surgery. J Neurosurg Spine 11:471–476

30. Wang MY, Lerner J, Lesko J, McGirt MJ (2012) Acute hospital costs after minimally invasive versus open lumbar interbody fusion: data from a US national database with 6106 patients. J Spinal Disord Tech 25:324–328

31. Park P, Foley KT (2008) Minimally invasive transforaminal lumbar interbody fusion with reduction of spondylolisthesis: technique and outcomes after a minimum of 2 years' follow-up. Neurosurg Focus 25:E16

32. Wang J, Zhou Y, Zhang ZF, Li CQ, Zheng WJ, Liu J (2010) Comparison of one-level minimally invasive and open transforaminal lumbar interbody fusion in degenerative and isthmic spondylolisthesis grades 1 and 2. Eur Spine J 19:1780–1784

33. Peng CWB, Yue WM, Poh SY, Yeo W, Tan SB (2009) Clinical and radiological outcomes of minimally invasive versus open transforaminal lumbar interbody fusion. Spine (Phila Pa 1976) 34:1385–1389

34. Weinstein JN, Lurie JD, Tosteson TD, Hanscom B, Tosteson ANA, Blood EA et al (2007) Surgical versus nonsurgical treatment for lumbar degenerative spondylolisthesis. N Engl J Med 356:2257–2270

35. Holly LT, Schwender JD, Rouben DP, Foley KT (2006) Minimally invasive transforaminal lumbar interbody fusion: indications, technique, and complications. Neurosurg Focus 20:E6

36. Schwender JD, Holly LT, Rouben DP, Foley KT (2005) Minimally invasive transforaminal lumbar interbody fusion (TLIF): technical feasibility and initial results. J Spinal Disord Tech 18(Suppl):S1–S6

37. Scheufler K-M, Dohmen H, Vougioukas VI (2007) Percutaneous transforaminal lumbar interbody fusion for the treatment of degenerative lumbar instability. Neurosurgery 60:203–212; discussion 212–3

38. Isaacs RE, Podichetty VK, Santiago P, Sandhu FA, Spears J, Kelly K et al (2005) Minimally invasive microendoscopy-assisted transforaminal lumbar interbody fusion with instrumentation. J Neurosurg Spine 3:98–105

39. Schizas C, Tzinieris N, Tsiridis E, Kosmopoulos V (2009) Minimally invasive versus open transforaminal lumbar interbody fusion: evaluating initial experience. Int Orthop 33:1683–1688

40. Shunwu F, Xing Z, Fengdong Z, Xiangqian F (2010) Minimally invasive transforaminal lumbar interbody fusion for the treatment of degenerative lumbar diseases. Spine (Phila Pa 1976) 35:1615–1620

41. Karikari IO, Isaacs RE (2010) Minimally invasive transforaminal lumbar interbody fusion: a review of techniques and outcomes. Spine (Phila Pa 1976) 35:S294–S301

42. Wu RH, Fraser JF, Härtl R (2010) Minimal access versus open transforaminal lumbar interbody fusion: meta-analysis of fusion rates. Spine (Phila Pa 1976) 35:2273–2281

43. Kim HJ, Jung WI, Chang BS, Lee CK, Kang KT (2017) And Yeom JS. Int J Med Robotics Comput Assist Surg 13:e1779

44. Sexton K, Johnson A, Gotsch A, Hussein AA, Cavuoto L, Guru KA (2018) Anticipation, teamwork and cognitive load: chasing efficiency during robot-assisted surgery. BMJ Qual Saf 27(2):148–154

45. Gillespie BM, Chaboyer W, Fairweather N (2012) Interruptions and miscommunications in surgery: an observational study. AORN J 95 (5):576–590

46. Kim H-J et al (2016) A prospective, randomized, controlled trial of robot-assisted vs freehand pedicle screw fixation in spine surgery. Int J Med Robot 13(3). https://doi.org/10.1002/rcs.1779

47. Bennett CR, Kelly BP (2013 Aug 9) Robotic application of a dynamic resultant force vector using real-time load-control: simulation of an ideal follower load on cadaveric L4-L5 segments. J Biomech 46(12):2087–2092

48. Kelly BP, Bennett CR (2013) Design and validation of a novel Cartesian biomechanical testing system with coordinated 6DOF real-time load control: application to the lumbar spine (L1-S, L4-L5). J Biomech 46(11):1948–1954

49. Shweikeh F, Amadio JP, Arnell M, Barnard ZR, Kim TT, Johnson JP, Drazin D (2014) Robotics and the spine: a review of current and ongoing applications. Neurosurg Focus 36(3):E10

50. Härtl R, Lam KS, Wang J, Korge A, Kandziora F, Audigé L (2013) Worldwide survey on the use of navigation in spine surgery. World Neurosurg 79:162–172

51. Ghasem A, Sharma A, Greif DN, Alam M, Maaieh MA (2018) The arrival of robotics in spine surgery: a review of the literature. Spine (Phila Pa 1976) 43(23):1670–1677

52. Overley SC, Cho SK, Mehta AI, Arnold PM (2017) Navigation and robotics in spinal surgery: where are we now? Neurosurgery 80(3S): S86–S99

Chapter 13

Robot-Assisted Anterior Lumbar Interbody Fusion

Patricia Zadnik Sullivan, William C. Welch, and John Y. K. Lee

Abstract

Robot-assisted anterior lumbar interbody fusion (rALIF) utilizes a robot for exposure of the anterior vertebral body to access the disk space for placement of a cage and plate for fusion of adjacent vertebral bodies. The goal of lumbar fusion is to limit painful, abnormal motion in the lumbar spine, and the assistance of robotic technology allows surgeons to reduce the incision size, length of stay, and postoperative pain. To date, a fully robotic ALIF is not feasible, as there are no FDA-approved bone-cutting instruments, including rongeurs and drills, that can be fitted to the robot. Limitations in surgical training and cost may further impact widespread use. The goal of this chapter is to review the benefits, indications, and limitations of the rALIF, and describe technical aspects of this emerging procedure.

Key words da Vinci, Anterior lumbar interbody fusion, Robot, Laparoscopy, Endoscope

1 Introduction

Lumbar fusion is indicated for patients with back pain due to lumbar spondylosis or spondylolisthesis, pars defects, or complete loss of disk height with resultant degeneration. Although there is debate in the literature regarding other indications, meta-analyses favor the use of lumbar fusion for a subset of patients with lumbar stenosis with or without spondyloarthropathy [1]. Posterior lumbar fusion may be offered with or without interbody fusion; however, the anterior lumbar interbody fusion (ALIF) procedure maintains the posterior tension band to prevent worsening spondylolisthesis. An ALIF procedure may also reduce certain risks, such as dural tears, associated with posterior lumbar fusions, however this benefit may come with increased overall cost and risk of readmission [2].

ALIFs require an abdominal approach, and manipulation of the abdominal viscera to access the ventral disk space. A traditional ALIF procedure involves exposure of a suprapubic or abdominal incision (determined by the level of the ALIF), followed by dissection for a transperitoneal or retroperitoneal exposure, and

Hani J. Marcus and Christopher J. Payne (eds.), *Neurosurgical Robotics*, Neuromethods, vol. 162,
https://doi.org/10.1007/978-1-0716-0993-4_13, © Springer Science+Business Media, LLC, part of Springer Nature 2021

placement of abdominal retractors to secure a safe surgical corridor to access the disk space. The correct level is confirmed using fluoroscopy. The disk material is incised using a scalpel and the disk is removed using pituitary and Kerrison rongeurs and bone curettes fixed to long handles. During this manipulation, the abdominal retractors may shift, and bowel or vascular structures may enter the surgical corridor. If neuromonitoring is in use, the muscular contraction of the rectus abdominus can further disrupt the abdominal retractors, complicating this approach, and putting nearby vascular structures at risk.

Historically, the risks with an open ALIF procedure include ileus, damage to abdominal viscera, and injury to the great vessels, in addition to hardware malposition or pseudoarthrosis [3, 4]. Use of an access surgeon can reduce but not eliminate these risks [5]. An open ALIF procedure involves a large abdominal incision, and proper closure is required to prevent ventral hernia formation. Like any open laparotomy, an open ALIF can be painful. For male patients, retrograde ejaculation is another risk that can result from transperitoneal manipulation of the presacral plexus [6].

Using a robotic assistant for the approach to the anterior spine, a surgeon can reduce pain from large incisions and mimic the benefits of laparoscopic general surgery, with an ideal outcome of decreased length of stay and improvement in postoperative pain. As the literature on rALIF is scarce, surgeons have extrapolated this improvement from various procedures within the general surgery and urological literature [7–9]. Further, use of a robotic assistant can ensure an improved view in complex patients, such as those who are obese.

2 Summary of Evidence

The IDEAL guidelines for research and innovation provide a structure for discussing novel technologies. The IDEAL acronym is short for idea, development, exploration, assessment and long-term monitoring. The idea for the robotic ALIF was initially proposed and piloted in a pig model, and transitioned to human cadavers, followed by human patients. The da Vinci surgical robot was adopted from general surgery and has not gained widespread use in the neurosurgical or orthopedic surgical communities. This is primarily due to the lack of instruments required for bone work associated with neurosurgical procedures. Specifically, rongeurs and other bone biting instruments have not been developed to attach to the da Vinci surgical arms.

The authors first described a robotic ALIF in a pig model in 2011 [10]. In this description, the authors used a 54 kg pig to assess the feasibility of the da Vinci robot for anterior disk exposure. The neurosurgeons in this group were given access to the da Vinci robot, however the authors described several collisions of the robotic arms, as well as freezing due to "unskilled motion." This required manual manipulation of the surgical arms prior to proceeding. The procedure took a total of 6 h. Beutler and colleagues later described the progressive implementation of the da Vinci system in a pig model, followed by a human cadaver and finally a human patient in 2013 [11]. In this description, the authors report using the da Vinci robot for access and transperitoneal exposure of the L5–S1 disk space in a female patient, followed by standard laparoscopy tools for disk removal and cage placement [11].

In the largest case series currently published, Lee and colleagues described the use of the robotic ALIF for 11 patients [12]. The authors described a subset of this cohort in an earlier technical note [13]. In this series, all patients demonstrated successful fusion and no cases were converted to the open approach. Of note, the average age in this series was 48.6 years and the average BMI was 26.5. A urologic surgeon used the da Vinci surgical robot to gain access to the ventral vertebral body via a transperitoneal approach. Fluoroscopy was obtained to confirm the correct level, and the remaining surgical procedure was conducted via laparoscopic approach by two neurosurgeons.

In this series, 5 of the 11 patients had a history of a prior abdominal surgery. Among the 11 patients, 7 patients underwent single level L4–5 or L5–S1 and 4 underwent two-level ALIF (L4–S1) and only 1 patient in this series demonstrated partial, asymptomatic cage extrusion. There was no incidence of bowel or vascular injury, and no reported cases of retrograde ejaculation among male patients.

Exploration and innovation are key in advancing novel surgical techniques, however the robotic ALIF has not gained widespread adoption. Only one case series has been published on this topic, and other groups should replicate this approach in order to determine the risks and benefits of the rALIF. Further, the da Vinci surgical robot is the only robot that has been utilized for the rALIF. As the field of robotic surgery advances, new robots should be tested to assess their utility in the rALIF. New centers are emerging to offer robotic assisted spinal surgery, and technological adjuncts such as intraoperative navigation have become widely used in spine surgery [14]. As new technologies emerge, the FDA will play a key role in long term monitoring for late and rare problems.

However, at this time, there are no FDA approved bone-cutting instruments that can be used for a fully robotic ALIF. A robot and access surgeon may be utilized for the initial approach, however there are no rongeurs, Kerrison, endplate shavers or other instruments that can be fitted to the robotic arms of the da Vinci robot. For this reason, the robotic ALIF is not truly a robotic approach, coupled with a laparoscopic ALIF. Gel ports can be used to maintain pneumoperitoneum while surgeons introduce laparoscopic rongeurs and endplate preparation tools under endoscopic guidance.

3   Description of Robot

In the literature, the da Vinci Surgical Robot (Intuitive Surgical, Sunnyvale, CA) has been described for the rALIF. Although models vary, the general conformation of the robot includes three or four arms that can be fitted with various surgical tools. The most common use for the da Vinci system is general, gynecological and urological surgeries, and tools have been designed for optimal use in these subspecialties. The robot has an operative component that is docked to the patient, fitted with the arms, and a console for the surgeon. The console is fitted with handheld wristed instruments that control the instruments at the tips of the robotic arms. This allows the surgeon increased degrees of freedom, including rotational movements, that would be impossible in a traditional laparoscopic approach. The use of the console also reduces certain elements of human error, especially hand tremors. The wristed instruments also have greater mobility than the human hand and wrist and can rotate more freely within the surgical corridor. The surgeon, seated at the console, watches a display of the surgical field from a seated position, and controls the instruments at the ends of the robotic arms. Models vary, and three-dimensional high-definition displays are available.

4   Description of Technique

Preoperatively, the access surgeon should be aware of the level of interest, and lateral and anteroposterior fluoroscopy may be used to localize the level of the incision. Neuromonitoring may be used to monitor nerve compression during the procedure. The patient is placed supine in Trendelenburg to facilitate retraction of the abdominal organs from the surgical field. Further, a modified Lithotomy position may be used as necessary to allow for docking of the da Vinci robot.

Laparoscopic access is gained using standard technique. Typically, the surgeon requires an entry port for the endoscope, as well as 12 mm ports for each of the tools. A grasper, cautery and or other tools may be required to mobilize the abdominal viscera to access the bony spinal column. These tools are fitted to the robotic arms. The robot is then docked to the patient, as the arms are aligned to the access ports and calibrated. A 30° endoscopic camera is also placed, typically via a supraumbilical port, to allow the surgeon a three-dimensional view inside the surgical cavity. In one case series, the authors describe the umbilical port for camera placement, as well as a far-right lateral assistant port, and two robot instrument ports [12]. In addition to these sites, a gel port site should be prepped into the field to allow the introduction of laparoscopic disk tools without loss of pneumoperitoneum.

The access surgeon, typically a general surgeon or urologist, then utilizes graspers and mono- and bipolar cautery to mobilize the bowel and access the spine. Of note, the level of the ALIF greatly influences the approach, as the L4–5 disk space is deep to the iliac bifurcation. For L4–5 ALIFs via a robot-assisted ALIF, red rubber catheters or other nontraumatic retraction devices may be required to safely access the disk space. Small lumbar branches of the aorta may be clipped using Weck clips (Teleflex Medical, Research Triangle Park, NC, USA) to avoid bleeding during later retraction. For L5–S1 ALIF approaches, the presacral plexus should be identified and carefully retracted. At this level, the sacral artery and vein may require ligation or clipping for a safe approach.

The anterior longitudinal ligament is incised, and fluoroscopy is obtained to confirm the correct level. At this time, the robot may be undocked from the patient, and the gel port (GelPOINT® Advanced Access Platform, Applied Medical, Rancho Santa Margarita, CA, USA) may be placed at a suprapubic site for the introduction of laparoscopic ALIF tools. The positioning of the gel port is determined by the angulation of the disk space of interest and should be tailored to each patient's needs.

At this stage, the disk space is then incised by the spinal surgeon and the disk material removed via pituitary and Kerrison rongeurs, as well as bone curettes. The endplates are then shaved with an endplate clearing device, and a sample cage is placed into the disk space. Serial cages may be fitted into the disk space until the ideal height and lordosis are achieved. Intraoperative fluoroscopy confirms placement, and the final cage is then placed and secured. For stand-alone cages, an awl may be used to drill pilot holes, or a plate may be placed on the anterior surface of the cage.

5 Tips and Tricks

Open ALIF approaches are traditionally described via retroperitoneal approaches. In male patients with L5–S1 ALIF, manipulation of the presacral plexus using transperitoneal approaches carries the risk of retrograde ejaculation [6, 15, 16]. For male patients wanting to father children, this can be a devastating complication and must be discussed prior to surgery. To date, there has been no description of a retroperitoneal robot-assisted laparoscopic ALIF, however the case series are too small to definitively determine the risk of retrograde ejaculation in robot-assisted ALIF with a transperitoneal approach.

In one case series, the authors report the use of a red rubber catheter as an atraumatic retractor for the great vessels when approaching L4–5 using robotic assistance [12]. This allowed retraction during placement of the disk spacers to reduce the risk of vascular injury.

6   Conclusion

Based on limited case series, the rALIF is a safe method for translating the benefits of laparoscopy to anterior spinal fusion. Limitations in widespread adoption include the lack of rongeurs and other spinal instrumentation for the da Vinci robot, as well as the overall cost of the robot. In order for the ALIF to be fully robotic, spine surgeons will also need to decide if robotic surgery training will be included in residency training, or if robotic surgery will require fellowship-level training. In human patients, training on the da Vinci robot is not standard for neurosurgical or orthopedic trainees, hence the use of an access surgeon in the reported human studies. Further, large-scale studies must be conducted to prove equivalence or superiority of the robotic approach to the mini-open ALIF or the traditional open ALIF.

References

1. Ahmed SI, Javed G, Bareeqa SB et al (2018) Comparison of decompression alone versus decompression with fusion for stenotic lumbar spine: a systematic review and meta-analysis. Cureus 10(8):e3135

2. Qureshi R, Puvanesarajah V, Jain A, Shimer AL, Shen FH, Hassanzadeh H (2017) A comparison of anterior and posterior lumbar interbody fusions: complications, readmissions, discharge dispositions, and costs. Spine (Phila Pa 1976) 42(24):1865–1870

3. Mobbs RJ, Phan K, Daly D, Rao PJ, Lennox A (2016) Approach-related complications of anterior lumbar interbody fusion: results of a combined spine and vascular surgical team. Global Spine J 6(2):147–154

4. Rothenfluh DA, Koenig M, Stokes OM, Behrbalk E, Boszczyk BM (2014) Access-related complications in anterior lumbar surgery in patients over 60 years of age. Eur Spine J 23(Suppl 1):S86–S92

5. Phan K, Xu J, Scherman DB, Rao PJ, Mobbs RJ (2017) Anterior lumbar interbody fusion with and without an "access surgeon": a systematic review and meta-analysis. Spine (Phila Pa 1976) 42(10):E592–E601

6. Lindley EM, McBeth ZL, Henry SE et al (2012) Retrograde ejaculation after anterior lumbar spine surgery. Spine (Phila Pa 1976) 37(20):1785–1789

7. Hajibandeh S, Sreh A, Khan A, Subar D, Jones L (2017) Laparoscopic versus open umbilical or paraumbilical hernia repair: a systematic review and meta-analysis. Hernia 21(6):905–916

8. Keus F, de Jong JA, Gooszen HG, van Laarhoven CJ (2006) Laparoscopic versus open cholecystectomy for patients with symptomatic cholecystolithiasis. Cochrane Database Syst Rev (4):Cd006231

9. Sajid MS, Bokhari SA, Mallick AS, Cheek E, Baig MK (2009) Laparoscopic versus open repair of incisional/ventral hernia: a meta-analysis. Am J Surg 197(1):64–72

10. Yang MS, Yoon DH, Kim KN et al (2011) Robot-assisted anterior lumbar interbody fusion in a swine model in vivo test of the da Vinci surgical-assisted spinal surgery system. Spine (Phila Pa 1976) 36(2):E139–E143

11. Beutler WJ, Peppelman WC Jr, DiMarco LA (2013) The da Vinci robotic surgical assisted anterior lumbar interbody fusion: technical

development and case report. Spine (Phila Pa 1976) 38(4):356–363

12. Lee Z, Lee JY, Welch WC, Eun D (2013) Technique and surgical outcomes of robot-assisted anterior lumbar interbody fusion. J Robot Surg 7(2):177–185

13. Lee JY, Bhowmick DA, Eun DD, Welch WC (2013) Minimally invasive, robot-assisted, anterior lumbar interbody fusion: a technical note. J Neurol Surg A Cent Eur Neurosurg 74(4):258–261

14. Overley SC, Cho SK, Mehta AI, Arnold PM (2017) Navigation and robotics in spinal surgery: where are we now? Neurosurgery 80 (3S):S86–S99

15. Carragee EJ, Mitsunaga KA, Hurwitz EL, Scuderi GJ (2011) Retrograde ejaculation after anterior lumbar interbody fusion using rhBMP-2: a cohort controlled study. Spine J 11(6):511–516

16. Comer GC, Smith MW, Hurwitz EL, Mitsunaga KA, Kessler R, Carragee EJ (2012) Retrograde ejaculation after anterior lumbar interbody fusion with and without bone morphogenetic protein-2 augmentation: a 10-year cohort controlled study. Spine J 12 (10):881–890

Chapter 14

Robot-Assisted Paravertebral Schwannoma Resection

Syed S. Razi, Nestor Villamizar, Michael Wang, John Paul G. Kolcun, and Dao M. Nguyen

Abstract

Robot-assisted thoracic surgery is at the forefront of minimally invasive approaches for resection of various intrathoracic benign and malignant tumors. The robotic approach is particularly suitable for solitary paravertebral schwannomas along the thoracic spine, lacking an intraspinal component, that are typically treated with complete resection and carry an excellent prognosis. Here, robotic platforms have established themselves as safe, reliable, and reproducible tools in expanding the surgeon's reach in complex tumor resections and enable junior surgeons to overcome the learning curve quicker without compromising oncologic principles or patient safety. This chapter appraises the existing literature and provides an in-depth review of the technical aspects of robot-assisted paravertebral schwannoma resection.

Key words Robotics, Schwannoma, Paravertebral, Neurosurgery, Spinal surgery, Robotic thoracoscopic surgery, Posterior mediastinal paravertebral neurogenic tumor

1 Introduction

Benign neurogenic tumors constitute the most common etiology of posterior mediastinal tumors, arising from thoracic nerve roots, sympathetic trunks, or intercostal nerves. The majority of neurogenic tumors (95%) are Schwannomas, which are benign tumor of the nerve sheath, composed entirely of Schwann cells. The size, location, and anatomical distribution of Schwannoma's vary, and can be intraspinal (including intra- and extradural components), extraspinal or a combination of both (dumbbell tumors). Schwannomas are most frequently located in the cervical spine, followed by thoracic and lumbar spine. These tumors are typically diagnosed incidentally on radiography (ultrasonography, CT or MRI), but can also present with neurogenic symptoms of pain, paresthesias, and/or motor weakness. These tumors have favorable prognosis as they are well encapsulated and hence are typically amenable to complete resection, although certain locations can present its own sets of challenges [1].

Hani J. Marcus and Christopher J. Payne (eds.), *Neurosurgical Robotics*, Neuromethods, vol. 162,
https://doi.org/10.1007/978-1-0716-0993-4_14, © Springer Science+Business Media, LLC, part of Springer Nature 2021

Resection of thoracic spinal/paraspinal schwannoma has typically been achieved anteriorly via transthoracic approach comprising of posterolateral thoracotomy or by a posterior approach via laminectomy with or without facetectomy and spinal instrumentation, or a combination of both approaches [2]. Minimally invasive video-assisted thoracic surgery (VATS), and in particular robot-assisted VATS—R-VATS (da Vinci Surgical System, Intuitive Surgical, Sunnyvale, CA) has been recently adopted for various intrathoracic procedures for both benign and malignant thoracic tumors. Several studies have shown minimally invasive intrathoracic procedures to be superior to open thoracotomy with significant reduction in perioperative morbidity, length of stay and quality of life with equivalent long-term oncologic outcomes [3–7]. The robotic platform offers the advantage of a three-dimensional high definition camera with $10\times$ optical magnification and three wristed instruments with seven-degrees of motion. These characteristics of the robotic platform makes it extremely suitable for posterior mediastinal tumor resection even in the most caudal or cranial extremes of the chest cavity.

2 Summary of Evidence

Numerous studies have already demonstrated the efficacy of robotic resection of anterior mediastinal masses, in particular thymectomies, in a safe and reproducible manner [6, 7]. The same principle and advantage of robotic platform has been adopted by our group and others to resect benign posterior mediastinal neurogenic tumors. Ruurda and colleagues first described the utility of R-VATS for resection of a posterior mediastinal tumor [8]. The authors utilized the standard left thoracoscopic approach for excision of a 4 cm schwannoma with one camera trocar, two robotic trocars, two assistant trocars for lung retraction and an additional assistant trocar for suctioning. Bodner and colleagues adopted the robotic system for excision of mediastinal tumors and published their early experience with three resections of paravertebral neuromas [9]. Many subsequent publications described the utility of the robotic approach for surgical resections of mediastinal tumors but only a few contain sufficient number of posterior mediastinal neurogenic tumors to be included in this review. Melfi and colleagues reported their single institution experience of 69 robotic resections for mediastinal lesions, 13 of which were resection of paravertebral neurogenic tumors (19%). There were no intra-operative complications and no operative mortality. The mean postoperative hospital stay was 4.3 days (range 3–10 days) [10]. Li and associates reported the first Chinese series of 167 consecutive cases of robotic mediastinal surgery, 17 of which were resection of schwannomas (10.2%). There was no operative mortality and three cases (1.8%) required conversion to sternotomy, none of which was for resection of posterior mediastinal neurogenic tumors. The mean postoperative hospital stay was 2.95 days (range 1–7 days) [11]. Unfortunately, neither series separately reported the perioperative outcomes of schwannoma resections. Between 6/30/2012 and 12/31/2018, we performed 95 mediastinal robotic procedures, 8 of which (8.4%) were to resect mediastinal schwannomas. There were four males and four females (age: 57.4 ± 11.5; mean $\pm$ standard deviation), robotic console time was 77.8 ± 14.0 min; length of hospital stay was 1.75 ± 1.16 days (range 1–4 days; median 1 day) and there was no 90-day morbidity. Cerfolio and associates described their early operative experience with robotic procedures for posterior or inferior mediastinal pathology including few cases of neurogenic tumors [3]. The authors provided detailed description of their learning curve and refinement of operative techniques for these mediastinal tumors, especially with regard to the earlier generation of robotic platform (da Vinci Si). With older models, the bulky external arms and limited maneuverability significantly impacts the instrument reach in the deep and narrow spaces of the posterior mediastinum. Additionally, the complex docking configuration of the older models precluded their widespread adoption in difficult to reach posterior mediastinal tumors. With advancements in the robotic platform and further refinement in techniques, the robot-assisted resection of posterior mediastinal tumors has been simplified for all the various locations of these paravertebral tumors. We have previously reported the ease of robot-assisted resection of posterior mediastinal schwannomas in both the cranial and caudal extremes of the thoracic paravertebral gutters, utilizing the same set of trocars in the seventh or eighth intercostal space for both tumors [12]. Figure 1 shows representative CT scan images (axial, sagittal and coronal views) of neurogenic tumors at different locations along the posterior mediastinum. There is ample evidence that robot assisted thoracoscopic resection of posterior mediastinal schwannomas is safe, effective, and reproducible with excellent perioperative outcomes.

3 Description of Robot

The Xi is the latest da Vinci surgical system approved for use in the thoracic cavity and its advanced platform addresses many of the limitations posed by the older models. In particular, the overhead architecture of the robot with rotating boom-mounted arms, extended instrument reach and targeting significantly overcomes the anatomic challenges posed by the posterior mediastinal tumors. The docking process is inherently simpler with its rotating boom that allows for complete access to the thoracic cavity while being docked from any position around the patient. This feature is

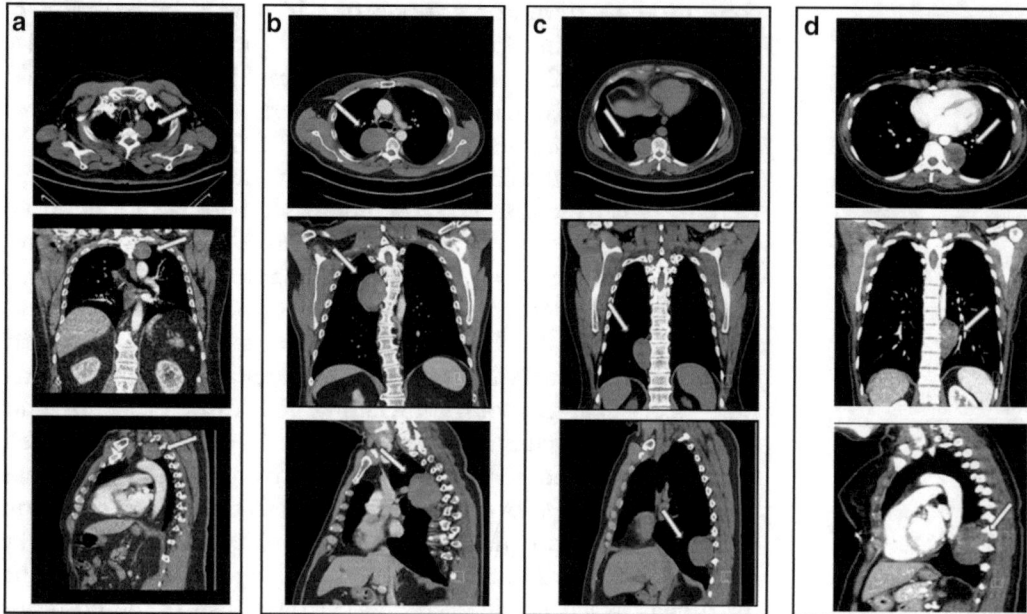

Fig. 1 Representative CT images (axial, coronal, and sagittal views) of neurogenic tumor at different locations along the posterior mediastinal space: left superior sulcus adjacent to the first rib (**a**); right inferior mediastinum abutting the 10th/11th vertebral bodies (**b**); a large right tumor abutting the tracheal carina (**c**); and a large left tumor in the low posterior mediastinum adjacent to the descending aorta (**d**)

particularly helpful with limited operating room spaces. The laser targeting feature during docking, and the use of simplified "linear" port configuration significantly reduces the docking time. A distinctive feature of the Xi is the redesigned 8-mm endoscope, that is able to be inserted through any of the 8-mm robotic ports, more commonly known as 'port hopping'. The working field for the four robotic arms is also significantly enhanced by making adjustments to the individual arm's patient clearance joints. These adjustments allow instruments from the same anatomic port configuration to reach both the cranial and caudal extremes of the thoracic cavity. The added benefit of the wristed staplers and ability to suture with ease provides a more comprehensive capability to the operating surgeon in posterior mediastinal tumor resections.

4   Description of Technique

General anesthesia is induced in a standard manner for thoracoscopy, and lung isolation achieved using an endobronchial balloon (Rusch EZ-Blocker, Teleflex Inc.). The patient is positioned in lateral decubitus and secured on a beanbag. The operating table is adequately flexed to widen the intercostal spaces of the operative side and to lower the hip from the operative field to allow for unhindered robotic arm movement. Additional safety straps and silk tapes are used to secure the position. The upper arm is rotated slightly cephalad and extended, hence allowing the scapula to rotate forward and providing greater access to the paraspinal spaces. Additionally, slight pronation of the operating table, allows the lung to drop away from the posterior mediastinal space. We use the seventh or eighth intercostal space for all port insertions, beginning with the camera port first, along the posterior axillary line. Carbon dioxide insufflation is used to deflate the lung, lower the diaphragm and gently displace the mediastinum to the contra-lateral side. Additionally, 8–10 mmHg intrathoracic pressure using CO_2 insufflation also confers hemostatic effects. One port anterior to the camera port, and two additional posterior ports are placed under the seventh rib and spaced approximately 8–11 cm apart. Finally, a 15-mm assistant port is placed, triangulated two rib spaces below the anterior and the camera port (Fig. 2). This port facilitates exchange of rolled gauze that assist in hemostasis in addition to serving as the port to extract the specimen. For posterior/inferior tumors located deep in the inferior sulcus, more complex port placement as shown in Fig. 3 is needed. Such arrangement is also frequently used for robotic transthoracic esophagectomy.

The robot is brought in approximately 60° to 90° to the operating table and the rotating boom mounted arms are aligned with the linear port configuration of the seventh intercostal space. The camera port is mounted, and after manually adjusting the camera to visualize the paravertebral tumor site, the automated targeting of the remaining arms is complete. This helps to maximize the work field and optimizes the movement of the arms centered around the targeted tumor. A Cadiere instrument is used for the left robotic arm, while a Maryland bipolar dissector is used for the right robotic arm. A tip-up grasper is typically used in the most posterior arm, allowing for atraumatic lung retraction.

Any pneumolysis needed is carried out to free up the adhesions between the lung and the chest wall. The parietal pleura overlying the tumor is grasped and incised using bipolar grasper. We use a "no-touch" technique focused on avoiding grasping the tumor at all times. This practice has prognostic implications for malignant tumors as it prevents tumor spillage, but even for benign masses it allows for maintaining the tissue planes around the tumor. Time and experience makes the dissection more efficient. The dissected parietal pleura overlying the tumor serves as a handle to retract the tumor and circumferentially dissect the adhesions around the tumor. Special attention is paid to expose the arterial branches coming off of the aorta, as undue traction on the tumor can unintentionally avulse these small branches and can lead to signifi-cant bleeding. With robotic surgery, it is absolutely critical to visualize the tissue tension lines developing with excessive traction. Bipolar energy is used to ligate small vessels, up to 2 mm. For larger

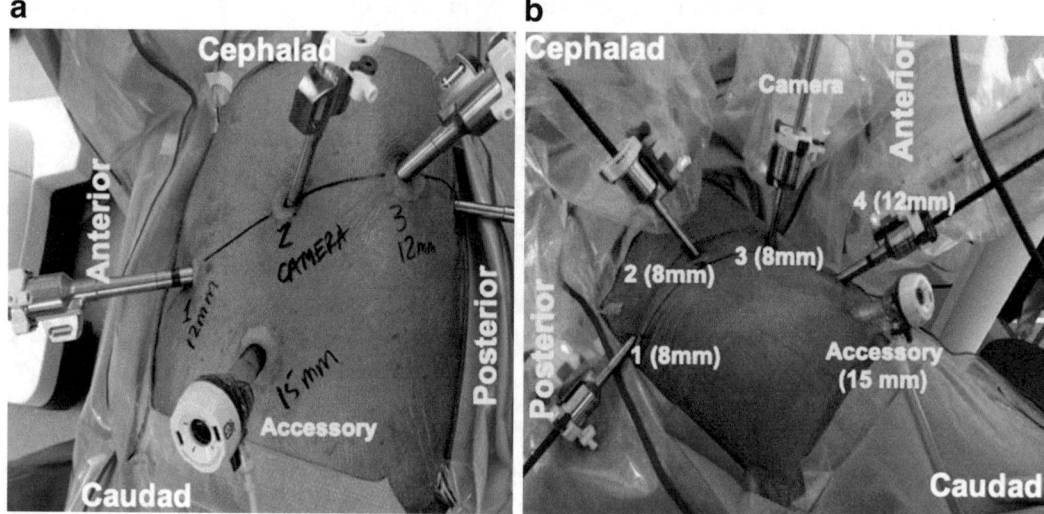

Fig. 2 Robotic port placements for robotic resection of posterior mediastinal tumors. (**a**) Left side: Ports 1–4 are spaced evenly (7–10 cm apart) along the seventh or eighth (for lower posterior mediastinal tumor in the inferior sulcus) intercostal space. # 1 is the most anterior 12 mm port fitted with a reducer to accommodate 8 mm instruments, # 4 is the most posterior port with the robot docked anteriorly at 60° to 90° to the body axis. (**b**) Right side: Ports 1–4 are spaced evenly (7–10 cm apart) along the seventh or eighth (for lower posterior mediastinal tumor in the inferior sulcus) intercostal space. # 1 is the most posterior port, # 4 is the most anterior 12 mm port fitted with a reducer to accommodate 8 mm instruments with the robot docked anteriorly at 60° to 90° to the body axis

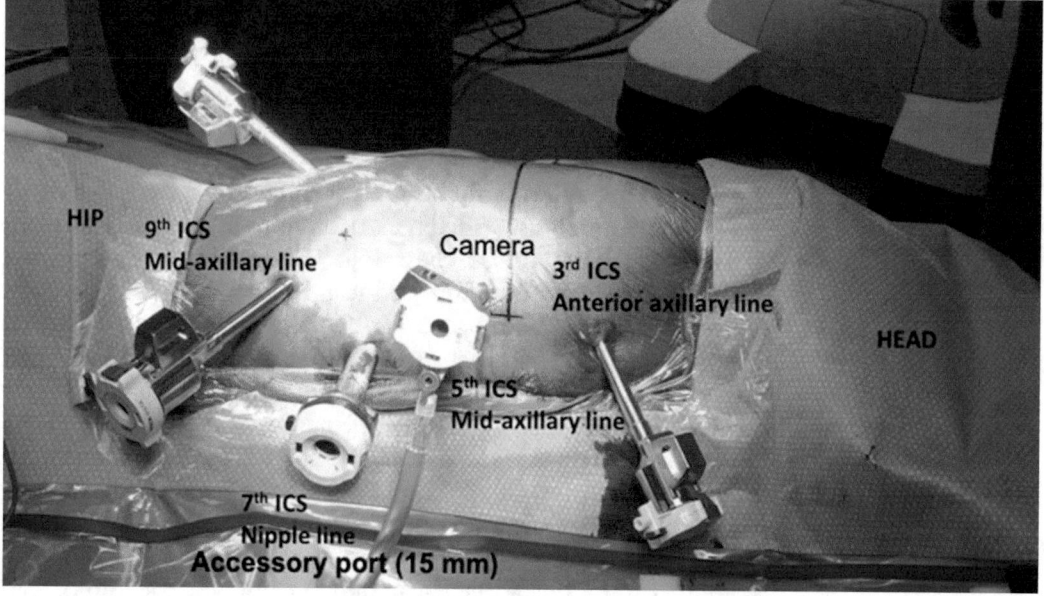

Fig. 3 Left robotic port placement for posterior mediastinal neurogenic tumor located deep in the inferior sulcus. Complex port arrangement is required for optimal view of the tumor and efficient movement of the working instruments

sized vessels, we recommend using the vessel sealer device. The mass is carefully dissected away from surrounding structures and removed with an intact capsule. Figure 4 shows representative intraoperative photographs of schwannomas in the left superior sulcus (a) or right inferior sulcus (b) with annotated surrounding anatomic landmarks pre- and postresection.

The specimen is retrieved in an endocatch bag through the assistant port. We routinely use intercostal nerve block with 1.3% liposomal bupivacaine. A 20-cc syringe with local anesthetic is connected to a 25-G winged catheter. Using the robotic arm, the winged tip of the 25-G needle is directed to infiltrate the second to eighth intercostal spaces with approximately 1.5 cc in each space. Meticulous hemostasis is achieved and verified. The de-docking maneuver is initiated with all robotics arms removed under direct vision and finally the port sites are assessed for any bleeding. A 28-French chest tube is inserted using the anterior port and secured along the paravertebral gutter while the lung is reinflated under vision. The remainder of the port sites are closed in standard two layered fashion.

The postoperative care for these patients is fast-tracked using "enhanced recovery after thoracic surgery" [ERATS] protocol. Pain control is optimized with scheduled pain medications. Oxycodone and morphine are administered only as needed. Chest tubes and Foley catheter are removed by early postoperative day 1, and most patients are discharged home on this day. Follow-up in the office is typically scheduled in 1–2 weeks post-operatively to check for wound healing. Patients are advised to gradually return to baseline function subsequently. The majority of posterior mediastinal tumors are benign neurogenic tumors, and with complete tumor excision, recurrence is rare and hence, routine surveillance is not needed.

5 Tips and Tricks

Intraoperative findings that would change management and mandate conversion to open thoracotomy include, but are not limited to, bleeding catastrophe and injury to the airway or esophagus. However, depending upon the experience and skill level of the operator, the latest da Vinci Xi surgical system does allow the surgeon to perform a plethora of reparative maneuvers to overcome these complications. The use of a bipolar grasper is particularly beneficial in dissecting the posterior mediastinal tumors, as it minimizes the tangential thermal spread of energy. Special attention is also paid to avoid unnecessary dissection in the posterior mediastinum on the right side to protect the thoracic duct. If the duct is identified and in close proximity to the tumor, it is safe to ligate it. If there is concern for thoracic duct injury, with accumulation of

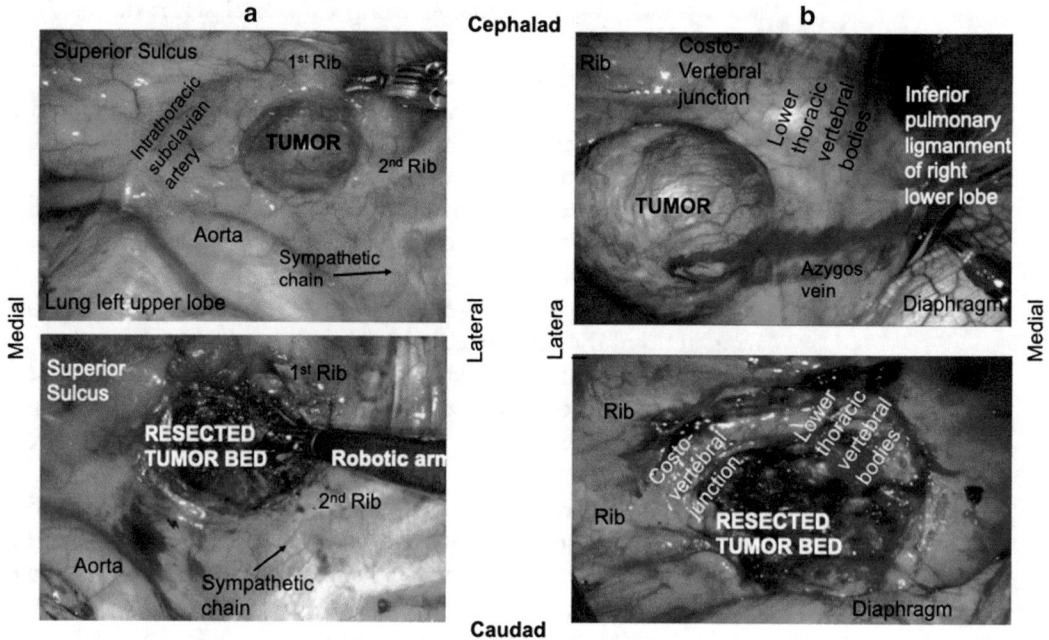

Fig. 4 Intraoperative images of pre- and postresection of posterior mediastinal neurogenic tumors by R-VATS (**a**: left superior sulcus neurogenic tumor; CT images are shown in Fig. 1a and **b**: right inferior posterior neurogenic tumor; CT images are shown in Fig. 1c) with annotated adjacent normal structure landmarks

chylous fluid, a mass ligation with a 2-0 silk suture is performed incorporating the entire mediastinal fat pad between the aorta (posteriorly) and the esophagus (anteriorly). Other important structures to clearly identify and avoid injury include the azygous vein, the sympathetic chain and the intervertebral foramina.

6   Conclusion

Robotic surgery has established its merit in various disciplines as a safe, reliable, and reproducible tool in expanding the surgeon's reach in complex tumor resections. While it may not offer the same benefit to a seasoned minimally invasive surgeon, it does certainly enable a junior surgeon to overcome the learning curve quicker and master more complex operations without compromising the oncologic principles or patient safety. Nonetheless, the cost-driven healthcare system does impose a significant challenge to successfully establish a robotic program. With advancements in the latest dV Xi surgical system, there is a tremendous potential to optimize operating times. The ease of docking and targeting significantly expands the reach of the operating arms and hence allows for more standardized approaches despite variations in anatomic locations of these posterior mediastinal tumors. Similarly, with further refinement in operative techniques, operative times, length of stay, and morbidity can be significantly reduced, allowing for more widespread adoption of robotic surgery.

References

1. Marchevsky AM (1999) Mediastinal tumors of peripheral nervous system origin. Semin Diagn Pathol 16:65–78

2. Ozawa H, Kokubun S, Aizawa T, Hoshikawa T, Kawahara C (2007) Spinal dumbbell tumors: an analysis of a series of 118 cases. J Neurosurg Spine 7:587–593

3. Cerfolio RJ, Bryant A, Minnich D (2012) Operative techniques in robotic thoracic surgery for inferior or posterior mediastinal pathology. J Thorac Cardiovasc Surg 143:1138–1143

4. Broussard BL, Wei B, Cerfolio RJ (2016) Robotic surgery for posterior mediastinal pathology. Ann Cardiothorac Surg 5(1):62–64

5. Yang MS, Kim KN, Yoon DH et al (2011) Robot-assisted resection of paraspinal schwannoma. J Korean Med Sci 26:150–153

6. Marulli G, Rea F, Melfi F et al (2012) Robot-aided thoracoscopic thymectomy for early-stage thymoma: a multicenter European study. J Thorac Cardiovasc Surg 144:1125–1132

7. Friedant AJ, Handorf EA, Su S et al (2016) Minimally invasive versus open thymectomy for thymic malignancies: systematic review and meta-analysis. J Thorac Oncol 11 (1):30–38

8. Ruurda JP, Hanlo PW, Hennipman A et al (2003) Robot-assisted thoracoscopic resection of a benign mediastinal neurogenic tumor: technical note. Neurosurgery 52:462–464

9. Bodner J, Wykypiel H, Greiner A et al (2004) Early experience with robot-assisted surgery for mediastinal masses. Ann Thorac Surg 78:259–266

10. Melfi F, Fanucchi O, Davini F et al (2012) Ten-year experience of mediastinal robotic surgery in a single referral centre. Eur J Cardiothorac Surg 41(4):847–851

11. Li H, Li J, Huang J et al (2018) Robotic-assisted mediastinal surgery: the first Chinese series of 167 consecutive cases. J Thorac Dis 10 (5):2876–2880

12. Pacchiarotti G, Wang M, Paul G et al (2017) Robotic paravertebral schwannoma resection at extreme locations of the thoracic cavity. Neurosurg Focus 42(5):E17

Chapter 15

Early Developments, Current Systems, and Future Directions

Taku Sugiyama, Sanju Lama, Hamidreza Hoshyarmanesh, Amir Baghdadi, and Garnette R. Sutherland

Abstract

Advances in neurosurgery have relied upon and often paralleled technological innovation. One of the latest technological breakthroughs to enter the neurosurgical operating room is robotics as a surgical adjunct. Although numerous types of surgical robotic systems have been designed and developed for various purposes, only a few systems reached commercial or clinical use, and none of them have been applied widely all over the world yet. This, in part, reflects the tardiness of the pull from the community that has attained a seemingly comfortable zone with advanced imaging, surgical navigation, and microsurgical technique. As science and engineering continue to advance, they allow increased freedom to explore the realms of microscopy, imaging, emerging cellular acoustics, leading-edge sensory feedback, teleoperative haptics, and their incorporation into the artificial intelligence–machine learning paradigm to further improve existing robotic technologies, toward automation. This review presents, in perspective, early development of neurosurgical robotics and current successful systems that have reached clinical trials, together with a vision for future direction in the development of neurosurgical robotic systems.

Key words Robot, Robotic surgery, Engineering, Technology, Neurosurgery, Microsurgery, Stereotaxy, Spine, Craniotomy

1 Introduction

Progress in neurosurgery has largely relied on and often paralleled advances in technology.

As the world of engineering and industry advanced toward robotics and automation in the middle of the twentieth century, investigators began to integrate robotic technology into surgery [1–4]. Robotic technologies are undoubtedly thought to be able to offer the potential for increased precision and accuracy, timewise stability, reduced tremor, motion scaling, and measurable sensory feedback. As medicine has become heavily dependent on advanced electromechanical sensing/actuation, supervisory-controlled instrumentation, hardware–software integration in data acquisition

Hani J. Marcus and Christopher J. Payne (eds.), *Neurosurgical Robotics*, Neuromethods, vol. 162,
https://doi.org/10.1007/978-1-0716-0993-4_15, © Springer Science+Business Media, LLC, part of Springer Nature 2021

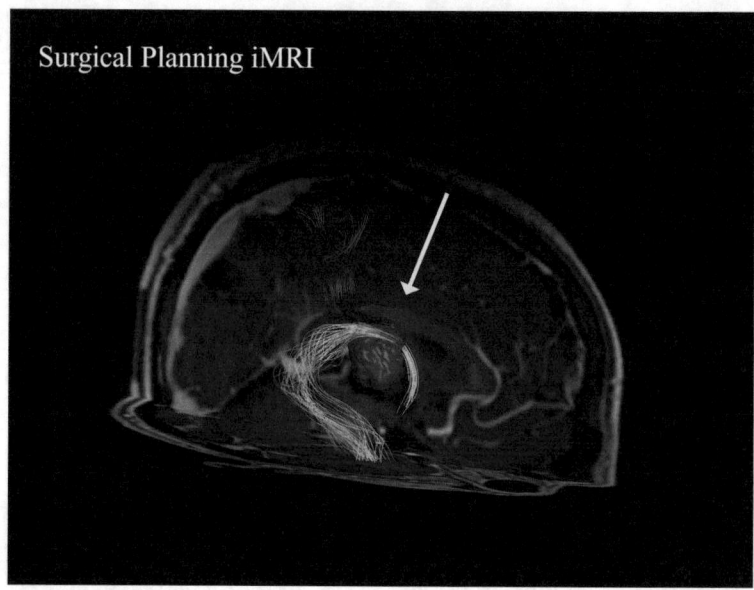

Fig. 1 Sagittal iMRI (T1 post gadolinium) with diffusion tensor imaging tractography, showing third ventricular tumor with relevant fiber tracts and planned interhemispheric surgical corridor (yellow arrow)

and display, imaging and lesion localization, the art of surgery combined with machine technology offers an exciting pathway to the future of surgery. This is particularly of interest to neurosurgery, where surgeons are intricately dependent on imaging and navigation technologies in 3D space (Fig. 1), and expanding field of molecular neuroscience guides and necessitates cellular visualization in pursuit of precision therapy.

Since robotic surgery first entered neurosurgery in mid-1980s, numerous surgical robotic systems have been conceptualized and designed for now over 30 years. However, only a handful of the systems have attained clinical use, and none of them have been widely accepted. In this review, we provide a comprehensive overview of various neurosurgical robotic systems, the nuances, advantages, and challenges, with an introspection on future directions. In a world where software and hardware technologies are advancing exponentially, and with med tech industries' competitive efforts, entry of robots into the neurosurgical operating rooms is both challenging and inevitable. Coupled to machine learning and automation, robotics will add advanced computer science and intelligence to the performance of surgery.

2 Definition and Classification of Robotics

The Robotic Institute of America defined the term *robot*, in 1979, as *a reprogrammable, multifunctional manipulator designed to move materials, parts, tools, or other specialized devices through various programmed motions for the performance of a variety of tasks* [5, 6]. Evolving toward a broader application, the current working definition of a robot is understood as a reprogrammable mechanical device used instead of or supervised by a person to perform sophisticated or repetitive tasks with a high degree of accuracy, repeatability, and stability. A robotic manipulator or a robotic arm is usually designed and constructed using rigid or nature-inspired flexible links connected by rigid or flexible joints having one end fixed to a frame, which is called *world frame*, and a free end that can move and perform tasks, that is called end-effector. Thereby, mechanically, robots could be classified into four types based on link–joint configuration: (1) rigid link–rigid joint, (2) rigid link–flexible joint, (3) flexible link–rigid joint, and (4) flexible link–flexible joint. In rigid links the ratio of the elastic deflection to the allowable applied load is negligible. Flexible links with inherent elastic or viscoelastic deformation are usually instrumented by a strain gage, measuring the flexural deflection, to be used as a state feedback in the control loop. Elastic printable programmable viscoelastic materials allow robot designers to take advantage of the properties of viscoelastic damping materials in order to reduce vibration and impact loads. Flexible joints, on the other hand, need harmonic drives similar to what is used in *Canadarm* and *Canadarm2*. As flexible links and joints are less common in surgical robotics, this review is limited to rigid manipulators.

3 General Engineering Characteristics

Displacement of the end-effector of the rigid manipulators relates to movement of joints. The accuracy and precision (repeatability) of the end-effector displacement is a function of material characteristics (density, stiffness, elasticity, creep, internal friction, and damping ratio) and postfabrication procedures (i.e., thermal treatment, finishing, lapping, honing, etc.). These may affect material deflection due to mechanical or thermomechanical stresses, together with design tolerances, manufacturing accuracy and precision. The terms *accuracy* and *precision* become meaningful when the robot end-effector is commanded to move to a designated position several times and the coordinate of those positions is measured/recorded each time in a 3D framework. How close/far the mean value of the recorded positions is to/from the desired commanded position determines the positional accuracy of the robot. The deviation of the recorded positions (disregarding the desired commanded target) also provides precision to robotics. Accordingly, a robot could be considered as high *accuracy–high* precision, high *accuracy–low* precision, low *accuracy–high* precision, and low *accuracy–low* precision. The workspace and intrinsic kinematic performance of such manipulators are ascribed to structural design and linkage dimensions that could be obtained through forward kinematics equations. Dynamic performance of the manipulator is also attributed to the mass and inertia of the links, gravity compensation as well as friction in the joints.

Robotic manipulators are classified based on different criteria such as *type of movement (DOF), application, brand, architecture,* and *collaboration capability.* From the DOF perspective, robots are classified as Cartesian robots, cylindrical robots, spherical or polar robots, articulated six-axis robots, SCARA robots, anthropomorphic robots, and redundant robots. Robots are also named based on the target applications as industrial and medical-grade robots (e.g., welding robots, material handling robots, palletizing robots, painting robots, grinding robots, assembling robots, and surgical robots).

Architecture refers to the kinematic design, according to which the robots are categorized into serial, parallel, and hybrid architectures. Serial linkage arms, made of multiple serially assembled links and joints, often provide a wide workspace, ease of motion, complicated trajectory paths, and high torque/force with significant manipulability, conditioning index, and dexterity. They also could mimic the human upper limb movements, which is a rewarding point for robot-assisted surgery. However, there are some limitations with the serial structures such as accumulated thermal and mechanical displacement errors, inertia, and gravity compensation. Parallel robots, taking advantage of multiple parallel pods, a fixed back plate, and a moving top plate, usually have short links, low inertia and provide high accuracy and precision. The pods could be prismatic or angular. The prismatic pods possess significant strength to generate high normal forces perpendicular to the fixed back plate. Angular pods normally consist of delicate spherical joints. Parallel robots have their specific limitation, too. They might face to overmobility or undermobility singularities in some portions of their rather limited workspace. They have stability limitations at high force scales, specifically when applying transverse forces in an inclined plane relative to the fixed back plate. Hybrid robots benefit from advantages of both serial and parallel structures to mitigate the limitations and challenges. From the collaboration aspect, robots are known as nonredundant, redundant, and highly redundant. A kinematically redundant manipulator possesses more degrees of freedom than it is needed to execute a given task, aiming to achieve more dexterous robot motions (less singularities). Most robot manipulators designed for surgical applications are articulated robots, as they provide more flexibility and degrees of freedom in movement.

---

4 Surgical Robots: A Perspective

As defined by the Society of America Gastrointestinal and Endoscopic Surgeons and Minimally Invasive Robotic Association (SAGES-MIRA) Robotic Surgery Consensus Group, *robotic surgery* is *a surgical procedure or technology that adds a computer-technology-enhanced device to the interaction between surgeon and the patient during a surgical operation, and assumes some degree of freedom of control heretofore completely reserved for the surgeon.* This definition encompasses micromanipulators, remotely controlled endoscopes, and console–manipulator devices. The key elements are enhancement of the surgeon's abilities—by vision, tissue manipulation, or tissue sensing—and alteration of the traditional direct local contact between the surgeon and patient [7]. It could well be stated that robotics provides a platform for augmented reality to the surgeon by interfacing between machine technology and human executive capacity [8].

There are multiple classifications of controlling robots based on their functional application and relationship between the robot and surgeon, as the field of robotics continues to evolve (Fig. 2) [9]. Surgical robotics may be broadly classified (in terms of who is in charge of controlling the robot motion) into three categories based on the fundamental principle of surgeon–machine interaction [10, 11]: (1) *Supervisory-control system*, in which the surgeon plans the operation, and the robot then carries it out autonomously under the supervision of the surgeon, (2) *Telesurgical (teleoperated or local–remote) system*, where the robotic end-effectors are remotely manipulated by the surgeon in real time, through a local workstation haptic interface, (3) *Shared-control system*, in which the surgeon and robot share control of the instrument installed on the robot end-effector, but the surgeon does most of the work. In such a system, the robot monitors the surgeon's performance and provides stability and support through active constraints and virtual fixturing to limit the operational workspace into the safe, close, boundary, and forbidden regions. Safe regions are considered the main focus of a surgery, defined by surgeons. In brain surgery, the safe region relates to the surgical corridor and lesion or target. The close region would be the boundary between the lesion and normal brain. The close region can be detected by provision of force feedback through a haptic interface, restricting the surgeon's hand motion as the surgical dissection approaches the boundary. Into this classification, an argument could be made for *Automated*

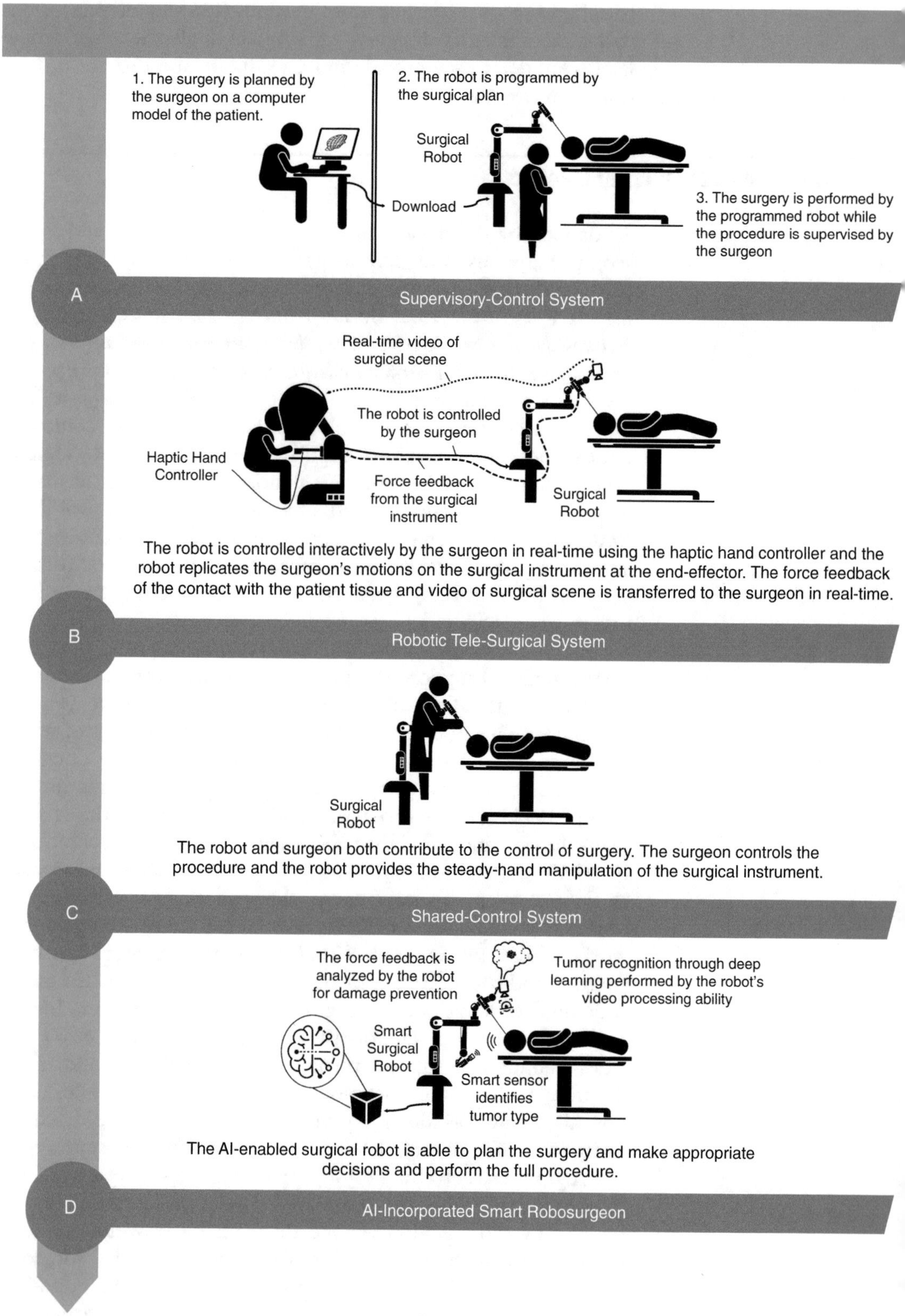

1. The surgery is planned by the surgeon on a computer model of the patient.

2. The robot is programmed by the surgical plan

Surgical Robot

Download

3. The surgery is performed by the programmed robot while the procedure is supervised by the surgeon

A   Supervisory-Control System

Real-time video of surgical scene

The robot is controlled by the surgeon

Haptic Hand Controller

Force feedback from the surgical instrument

Surgical Robot

The robot is controlled interactively by the surgeon in real-time using the haptic hand controller and the robot replicates the surgeon's motions on the surgical instrument at the end-effector. The force feedback of the contact with the patient tissue and video of surgical scene is transferred to the surgeon in real-time.

B   Robotic Tele-Surgical System

Surgical Robot

The robot and surgeon both contribute to the control of surgery. The surgeon controls the procedure and the robot provides the steady-hand manipulation of the surgical instrument.

C   Shared-Control System

The force feedback is analyzed by the robot for damage prevention

Tumor recognition through deep learning performed by the robot's video processing ability

Smart Surgical Robot

Smart sensor identifies tumor type

The AI-enabled surgical robot is able to plan the surgery and make appropriate decisions and perform the full procedure.

D   AI-Incorporated Smart Robosurgeon

Fig. 2 Classification of surgical robotic system based on functional application

robotics as the fourth and ultimate category, where machine learning replaces the executive input from the surgeon, creating a smart, efficient, and independent surgical robot.

5 Differences Between Surgical Robots and General-Purpose or Industrial Robots

Similar to space robotics or robots designed for military applications, there are some extra precautions, factors, and considerations that need to be addressed in design and development of neurosurgical robots to provide a high level of fidelity, stability, and safety. Design considerations include ease of motion (mimic the task performed manually by surgeons), reasonable size, weight, inertia, and appearance, high manipulability, vibration damping, dexterous operational workspace, sensing network, accurate and precise trajectory planning, obstacle avoidance, virtual fixtures, stability and reliability as well as robust driving and actuation system. Stability, itself, covers a broad definition including *mechanical stability* in terms of vibration amplitude, frequency and damping ratio, *control stability*, that is, how quickly a controlled signal/present-value can follow the signal/value set by the user, and *electrical stability* which means the response of a circuit to the uncertainties and noise interferences. Reliability also reflects the concept of no significant hardware/software changes over time. These considerations are usually studied in the preliminary design phase of robot development and preparation of design requirement documents, as it causes many changes in the hardware design, manufacturing, assembling, and control processes that usually cannot be changed once robot development is completed. Software and control considerations in this regard can usually be changed after hardware development; however, they need to be considered during the design phase early on, as any untoward hardware design errors or ignorance could pose considerable challenge for the already developed control system or software, to compensate for after the fact. These compensations might even compromise other robot characteristics such as precision, accuracy, stability, efficiency, responsivity, and safety. The safety of a surgical robot comes to a vital priority (Fig. 3). A redundant sensing system including magnetic rotary encoders, optical rotary encoders, magnetic linear scales, optical linear scales, optocouplers (optoisolators), Hall-effect sensors, proximity inductive and capacitive sensors, remarkably enhances the safety of robotic surgery. The robustness of the driving and actuation systems has also been addressed by employment of ball-driven power transmission mechanisms such as ball screws and linear motion guides coupled with piezoelectric linear motors. To conclude, utilizing an industrial robot for medical applications is not the best practice, and a robot is better to be designed based on the targeted application in mind. As there are always tradeoffs in the

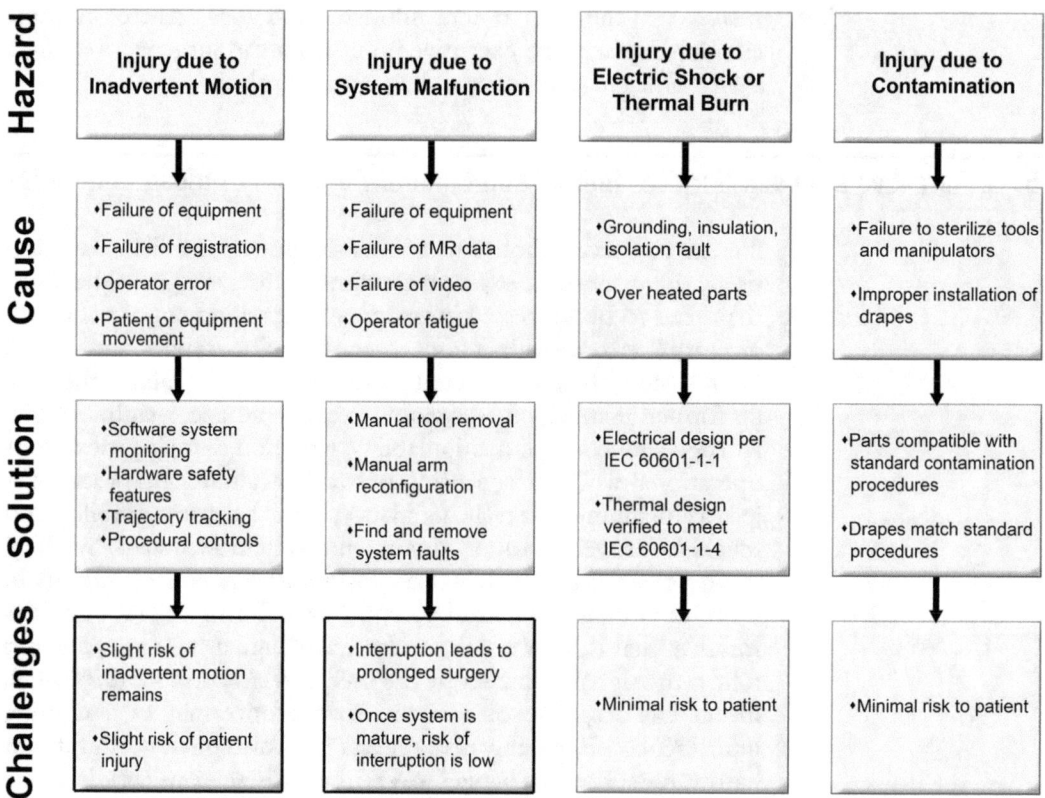

Fig. 3 Hazards related to the introduction of robotics into surgery. Described are the causes of the hazard, the control used to mitigate the risk, and the remaining challenges related to the hazard

engineering design process, designing a robot for too many applications *will* result in a robot that is rather deficient of each of those applications, and incurring increased cost than it would if designing multiple robots, each for one particular application.

6   Neurosurgical Robotics: Early Development and Current Systems

In this section, the review describes robotic systems and projects that reached commercial availability and/or clinical use (Table 1). The systems that were designed for other surgical discipline but evaluated in neurosurgery have also been included. Robotic platforms for endovascular intervention have not been included.

6.1   Robotics for Stereotaxy

The first robot introduced in neurosurgery was the Programmable Universal Machine for Assembly (PUMA; Advance Research and Robotics, Oxford, Connecticut), an industrial robotic arm with six DOFs and modified for stereotactic neurosurgery. By interfacing this system with a CT scanner and addition of a probe guide on its arm, Kwoh et al. used the PUMA robot for biopsy needle insertion

Table 1
List of successful robotic project for neurosurgery

Project name	Producer	Type	Feature	Regulation approval
Stereotaxy for brain				
NeuroMate	Renishaw Mayfield, Nyon, Switzerland	Supervisory	Mobile base, IEE with five DOF, frame/frameless US and CT-based registration	CE, FDA
ROSA	Medtech, Montpellier, France	Supervisory, Shared-control	Mobile base, IEE with six DOF, frame/frameless registration, haptic ability	CE, FDA
SurgiScope	ISIS Robotics, Saint-Martin-d'Hères, France	Supervisory	Ceiling-mounted seven DOF robotized manipulator with 2 modes: microscope and biopsy modes	CE, FDA
PathFinder	Prosurgics/Armstrong Healthcare Ltd. Coleraine, UK	Supervisory	Base attached to head clamp, IEE (guided needle and guided drill) with six DOF, frameless registration using fiducial markers	FDA, discontinued
CAS-R-2/NeuroMaster	Beijing University/Navy General Hospital, China	Supervisory	Mobile base, IEE with six DOF, frameless registration using fiducial markers	
PUMA	Advances Research and Robotics, UK	Supervisory	IEE with six DOF, industrial robort first used in neurosurgery	Discontinued
Minerv	University of Lausanne, Switzerland	Supervisory	IEE with five DOF, dedicated CT scanner	Discontinued
MKM	Carl Zeiss, Oberkochen, Germany	Supervisory	An operative microscoep mounted on six DOF arm for microscope navigation and and tool guidance	Discontinued

(continued)

Table 1
(continued)

Project name	Producer	Type	Feature	Regulation approval
Stereotaxy for spine				
Renaissance/ SpineAssist	Mazor Robotics Ltd., Caesarea, Israel	Supervisory	Portable hexapod parallel robotic manipulator with six DOF, directly mounted on a bony element of the surgical field	CE, FDA
iSYS 1	iSYS Medizintechnik GmbH, Kitzbühel, Austria	Supervisory	Miniature robot with four DOF, a passive seven DOF positioning arm used for prepositioning of robot	(CE, FDA)
AcuBot	John Hopkins University, Baltimore, Maryland	Supervisory	six DOF robot arm, a passive seven DOF positioning arm used for prepositioning of robot	(FDA)
Innomotion	Innomedic, Herxheim, FZK Karlsruhe, TH Gelsenkirchen, Germany	Supervisory	six DOF pneumatically driven robot arm, firm attachment of the system to the table	(CE)
Laser ablation				
NeuroBlate	Monteris Medical Inc. Plymouth, MN, US	Telesurgical	A laser device with two DOF, MRI compatible	FDA
Radiosurgery				
CyberKnife	Accuray Inc., Sunnyvale, CA	Automated	Six DOF robotic arm uses a guidance system to track the location of tumors in real-time and automatically adjust its focus with frameless system	CE, FDA, Japan
Novalis	BrainLab AG, Munich, Germany	Automated	Three imaging modalities for pinpointing the tumor and positioning the patient with high precision	CE, FDA, Japan

Craniotomy

Name	Institution	Control	Description	Status
RobaCKa	Karlsruhe University/ Heidelberg University	Supervisory	Six DOF robot, force-/torque-sensor, infrared navigation system, and an industrial PC collecting and processing sensor information	Discontinued

Endoscopic surgery

Name	Institution	Control	Description	Status
Evolution-1	Universal Robot Systems, Schwerin, Germany	Telesurgical	Mobile base, four DOF hexpod robot with high accuracy and great payload capacity	Discontinued
NeuRobot	Shinshu University, Matsumoto, Japan	Telesurgical	An End effector (10 mm diameter) consists of rigid 3D endoscope, 2 forceps and a laser, six DOF (3 at the micromanipulators)	Discontinued
da Vinci	Intuitive Surgical, Sunnyvale, California	Telesurgical	Two articulated arms with seven DOF, a separate camera arm caries a 3D endoscope	(FDA, CE)

Microsurgery

Name	Institution	Control	Description	Status
EXPERT	Shinshu University, Matsumoto, Japan	Shared-control	Arm holder that acts as a passive controlled robot with five DOF	Japan
neuroArm/ SYMBIS	IMRIS Inc., Minnetonka, MN, US	Telesurgical	Mobile base, 2 arms with seven DOF, MRI compatible, haptic feedback, motion scaling with tremor filtering	FDA (510k)

into the brain in 1985 [12]. Later, investigators in Toronto Canada reported successful application of the PUMA 200 robot as a retractor holder in the resections of thalamic astrocytomas in six children [13]. However, because of the safety concerns related to the use of an industrial robot in surgery, further development of PUMA for applications to neurosurgery was discontinued. It is also conceivable that introducing a seemingly complex and intimidating technology into the operating room, with no clear advantage over existing methods and art of performing surgery, also may have played a role in its abandonment.

In 1987, NeuroMate (Integrated Surgical Systems, Davis, California) was developed specifically for neurosurgery, with the intention for stereotaxy under image-guidance (Fig. 4) [14]. The system has a single arm with five DOFs and is capable of gripping or stabilizing surgical tools such as cannulas and biopsy needles. Using preoperative CT or MR brain images, the target lesion and trajectory are defined at a computer workstation, registering the robot arm to the predefined target. The recent version of this system (Renishaw Mayfield, Lyon, France) became the first Food and Drug Administration (FDA)-approved robotic system for neurosurgery. Although its bulky architecture is thought to be a negative feature, this system now ~60 installations worldwide and has been used in over 10,000 neurosurgical procedures, including deep brain stimulation (DBS) [15–17], stereoelectroencephalography [18–20], endoscopic approach [21], and stereotactic guidance to brainstem and convection-enhanced delivery of chemotherapy [22, 23].

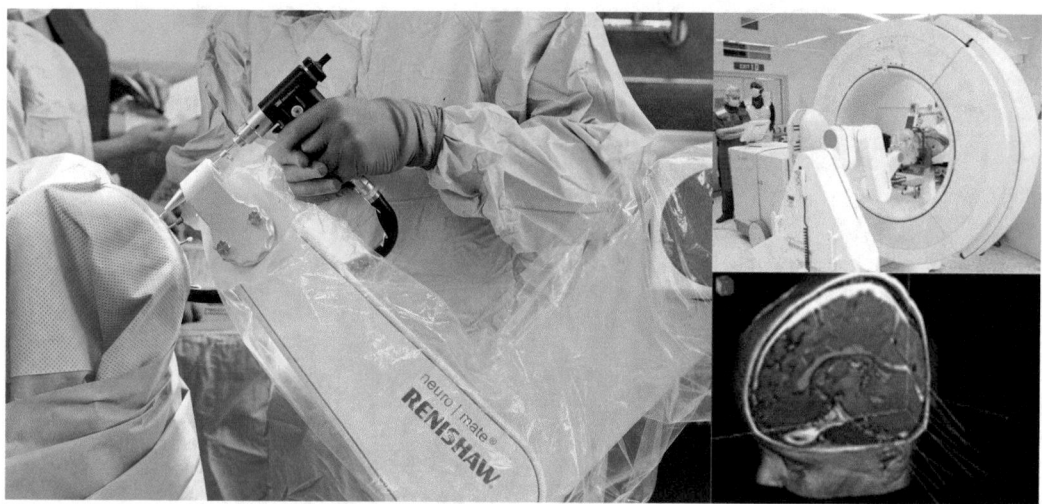

Fig. 4 The Neuromate System performing stereotactic electrode placement (inset: able to work within an imaging system for real time neurophysiologic stimulation and recording (top)); Reconstructed MR image with coordinates for target localization

The PathFinder Robot (Prosurgics/Armstrong Healthcare Ltd. Coleraine, UK) is a similar system with a guided needle for biopsy and drill for burr hole placement [24]. The characteristics of this system is that it uses camera system and identified reflectors attached to the patients' head instead of radiological imaging to specify a target and trajectory. The system has high accuracy with consistency in all targets in a patient, with less than 1 mm of mean deviation from the original plan [25]. It was also approved by FDA for neurosurgical applications in 2004, and tested clinically for epilepsy surgery [25, 26]; however, it was discontinued due to a lack of commercial interest, perhaps related to the widespread use of *stand-alone* navigation technology for lesion localization which was reliable, widely accepted, and relatively inexpensive.

Minerva (University of Lausanne, Switzerland), a robot with a five-DOF arm, was designed to perform stereotactic neurosurgical procedures within the CT scanner [27]. Minerva occupied a position at the head of the operating table, behind the CT scanner, so that its arm could work within the gantry; thus, the system demonstrated the first real-time use of CT guidance for stereotactic brain biopsy and overcame the problem of brain shift during surgery [27]. Although the integration of real-time imaging with robot system was an important concept and of interest, the project was discontinued because of safety issues. Other disadvantages include the need for a dedicated CT scanner, and again, due to competition with the *stand-alone* navigation systems widely being used in neurosurgery.

Concurrent advancement of near-real-time image guidance technology target localization and the development of frameless stereotaxy appeared to slow down the pace of evolving robotics for neurosurgery. On the contrary, technologies such as CyberKnife (Accuray Inc., FDA approved in 1999) [28, 29] and Novalis (BrainLab Inc. and Varian Medical Systems, FDA approval in 2000) [30] gained popularity in their use in radiosurgery. For CyberKnife alone, there are now over 240 systems installed worldwide with over 100,000 patients treated. Following this principle, the NeuroMate is also able to be used in a frameless mode and can be combined with a unique ultrasound registration system. The ROSA (Medtech SAS, Montpellier, France, now under Zimmer Biomet USA) is the latest generation of the neurosurgical robot that has been gaining acceptance for frameless stereotactic neurosurgical procedures (Fig. 5). The system is simultaneously a supervisory-control and a shared-control robotic device, that consists of a six-DOF manipulator and control software for planning, registration and guidance. This system has over 70 installations worldwide, with over 4000 surgical interventions including brain tumor biopsies, DBS, electrode placement [31–34], and laser ablation procedures [35–37]. Of interest, the ROSA Spine version has also been tested clinically [38–40], and received US FDA approval and the CE mark recently [41].

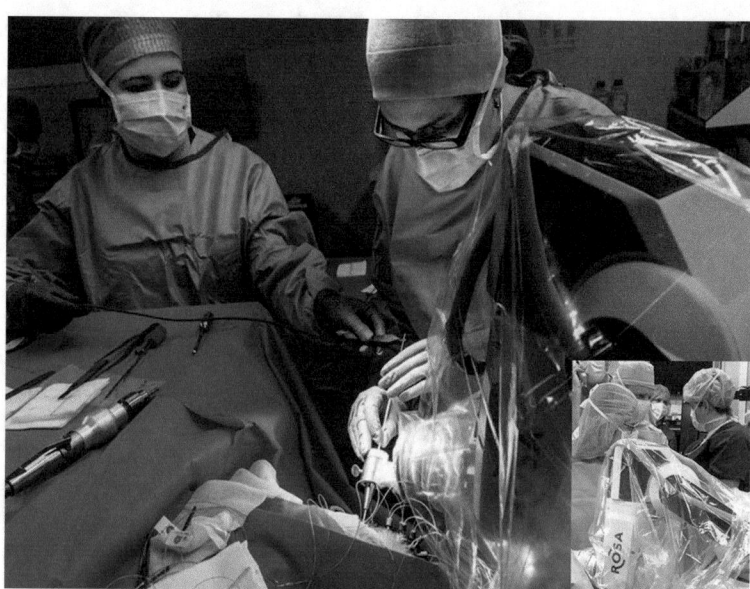

Fig. 5 The ROSA Robotic system assisting with stereotactic electrode placement (inset: the system integrated to operating room workflow)

6.2 Robotized Microscope

The Mehrkoordinaten Manipulator (MKM) systems (Carl Zeiss, Oberkochen, Germany) consists of an operating microscope that is mounted on a six-DOF motor driven robot and computer workstation. Its initial goal was to serve as a frameless stereotactic navigation system that combined the concepts of intraoperative microscopy and neuronavigation for minimally invasive surgery. The microscope is able to move independently to any position predefined by navigational planning [42, 43], and a biopsy needle holder could be attached to the bottom of microscope [44]. This system was clinically tested for frameless stereotaxy in several pathologies with acceptable results [42–47]; however, the project was discontinued due to its bulky architecture and high cost, which played an important role in its demise from market uptake. More recently, autopositioning robotic microscope was developed in collaboration with a neuronavigation company (Medtronic Navigation, Inc., Louisville, Colorado) and Carl Zeiss [48].

The SurgiScope system (ISIS Intelligent Surgical Instruments & Systems, Grenoble, France), a ceiling-mounted robotized microscope with seven-DOF arms, also has dual modes: microscopic and biopsy. In microscopic mode, it serves as platform for a microscope. Once a trajectory is defined, the focal point of microscope can be brought to the prescribed target. In biopsy mode, a biopsy arm with probe holder is attached to the bottom of the microscope, and the robot is activated to move the arm and probe holder to a prescribed trajectory [49]. The first system was installed in the operating room in 1996, and since in over 40 operating rooms

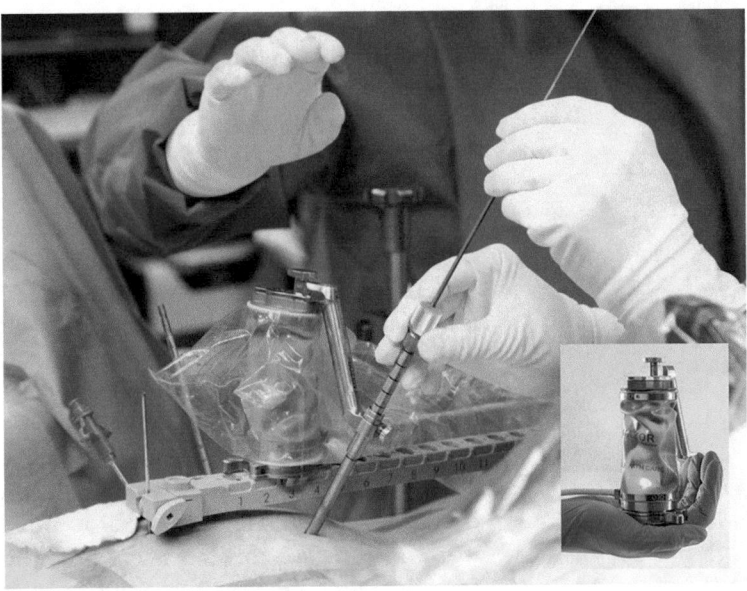

Fig. 6 The SpineAssist robotic system (inset: relative small size of the robot, often likened to that of a Coca-Cola can)

worldwide. Again, the system has similar drawbacks as the MKM systems. SurgiScope has been clinically used for frameless stereotactic brain biopsy, catheter, reservoir or electrode placement with high feasibility and accuracy [49–54].

6.3 Robotics for Stereotactic Spine Surgery

With the majority of robotic projects for spinal instrumentation beginning after 2000, the SpineAssist/Renaissance (Mazor Robotics Ltd., Caesarea, Israel, now Medtronic USA) became the first FDA-approved robotic system for spine surgery (Fig. 6) [55, 56]. Its main innovation and advantage was the reduced size and weight (base diameter 5 cm × height 8 cm, 250 g weight) so that it could be directly attached to the patient's bony structure, enabling and simplifying the registration process of imaging study. For open procedure, the robot can be directly mounted over spine using clamp and bridge. The system can be also mounted on a plastic railing anchored on two points of patient's pelvis and one of the spinous processes of an upper vertebra. To date, SpineAssist has been validated by >11,000 cases worldwide [57–66], and is only the system which has been used in comparative study with conventional freehand procedures in neurosurgery [67–79]. According to a meta-analysis of randomized controlled trials [75], there were no differences in the accuracy between robot-assisted and conventional freehand pedicle screw placement in both percutaneous and open robot-assisted methods; however, recent studies indicated that robot-assisted methods was associated with reduction of radiation exposure, short patient stay, and fewer proximal facet joint violations [72, 73]. Recently, this system was also tested for brain biopsy with a small platform connected directly to the patient's skull [80].

AcuBot (Hopkins/Georgetown, USA) [81], INNOMOTION (Innomedic/FZK/FH, Germany) [82], and iSyS1 (iSyS Medizin-technik/ACMIT, Austria) [83, 84] were commercially available systems designed for needle-guided interventions for abdominal organs rather than central nervous system. However, the AcuBot and INNOMOTION were also utilized for spinal nerve blockade and diagnostic needle biopsies [81, 82]. The iSyS1 system was tested for brain tumor biopsy and implantation of depth electrode in refractory epilepsy patients [83, 84]. The system is now acquired by Medtronic USA and marketed as AutoGuide.

6.4 Robotics in Telemedicine

From early 1960s, telemedicine has gradually progressed to become an integral aspect in healthcare. In this context, *telesurgery* using robotic systems has been attempted with the first reported case that described a cholecystectomy being performed between New York and Strasburg, spanning a distance of about 4000 km [85]. In neurosurgery, telementoring using robotic endoscopic system (AESOP, Computer Motion Inc., Santa Barbara, CA) that could be controlled remotely was also attempted [86]. In China, the distribution of the medical facilities allocates mainly to large cities such as Beijing and Shanghai, although a large fraction of population is located in rural areas. Therefore, telemedicine techniques have been developed since the late 1980s. In this context, telesurgical robotic systems were developed such as the Neuro-Master and the CAS-R-2 system (Beijing and Navy General Hospital, Beijing, China). The first case of keyhole stereotactic brain biopsy was conducted in 2003 between Beijing and Yan'an with a distance of about 1300 km [87, 88]. It is reported that >2000 cases of surgeries have been conducted out since then [89–93].

6.5 Robotics for Laser Ablation

The NeuroBlate system (Monteris Medical Corp., Plymouth, MN, USA) is a specifically designed telesurgical robotic system for MRI-guided ablation therapy [94]. It is designed to permit the neurosurgeon to remotely operate the two-DOF robotic device and deploy a laser to ablate brain tumors. The NeuroBlate device mounts into a separate, passive, stereotactic frame that is compatible with and visible to the MRI machine so that the surgeon can monitor the thermal dose using real time MRI thermometry data during procedure. A phase I clinical study was performed successfully [95], and received FDA approval in 2009 without specific limitations for intracranial use. The system was used for the cases of glioblastoma, metastatic tumor [95–98], and radiation necrosis [99, 100].

6.6 Robotics for Craniotomy

Craniotomy is still a challenging surgical task for robot kinematics because they normally allow for movement of the tool within a large volume but with constraint on orientations [101, 102]. In 2003, a patient with a lesion in the petrous bone was treated with the RobaCKa system (Heidelberg University/Karlsruhe University, Germany) [102, 103] which was based on a former industrial robot; however, the project was discontinued. Recently, handheld robotic systems are being developed for this purpose such as Craniostar (University of Heidelberg, Heidelberg, Germany) and Safe Trepanation System (RWTH-Aachen, Aachen, Germany) [104, 105].

6.7 Robotics for Endoscopic Neurosurgery

The Evolution 1 robot (Universal Robot Systems, Schwerin, Germany) was a hexapod platform that allowed precise movements in all six DOFs designed for endoscopic neurosurgery [106]. The platform is prepared to accept different types of surgical instruments via connection to a universal adapter and has a positioning accuracy of 20 μm. The motion resolution is 1 μm, even with loads of up to 500 N. This system was used in third ventriculostomy, fenestration of cystic lesion, pellucidotomy, and transsphenoidal removal of pituitary adenoma [106–108]. However, such a high payload capacity is not necessary for neurosurgery, and a parallel actuator limited its flexibility such that the project was discontinued. Furthermore, it is possible that the end-users who tested the system deemed it too bulky and time-consuming to justify use.

NeuRobot (Shinshu University School of Medicine, Matsumoto, Japan) was the telesurgical robot designed specifically for keyhole microneurosurgical and microendoscopic surgery [109]. NeuRobot has one long, thin manipulator arm (10 mm diameter), which houses a rigid 3D endoscope and three micromanipulators (1 mm forceps). Each micromanipulator has three DOF and has a corresponding hand controller at the workstation. NeuRobot has been used successfully in third ventriculostomy and partial removal of meningioma [109–111]; however, the project discontinued because it did not reach enough quality of dexterity for microsurgical procedure compared to human hand [112]. Other disadvantages were lack of sensory feedback and cost of the robot, which made it less attractive to investors.

The da Vinci surgical system (Intuitive Surgical, Sunnyvale, California) is the most frequently used telesurgical robot in a wide range of surgical procedures particularly within the field of urology and Ob-Gyn (Fig. 7). A major advantage of the da Vinci system is the ergonomy and ease-of-use provided by the anthropomorphic master console that restores the motor–visual alignment of the camera and surgical instruments. Recently, several groups have demonstrated the feasibility of using the da Vinci system in an anecdotal case reports in resection of paraspinal lesion [113–115], transoral odontoidectomy [116], and assistive use in anterior

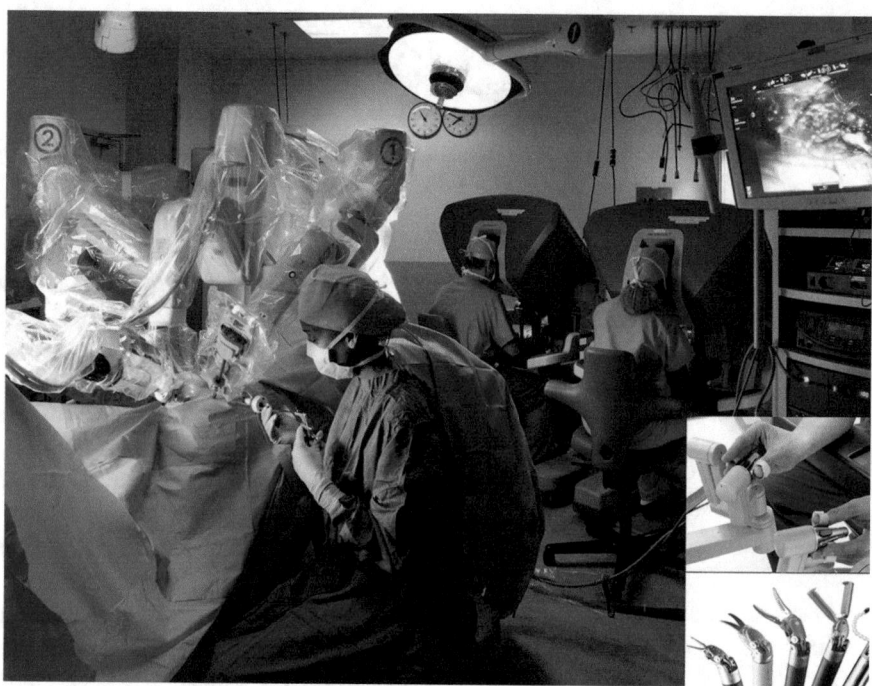

Fig. 7 The da Vinci Surgical system (inset: intuitive hand piece—hand controller at the human–machine interface (top), array of task specific toolset (bottom))

lumbar body fusion [117, 118]. The system were also tested in cadaveric study for keyhole neurosurgery [119, 120]; however, this system is not well suited to neurosurgical keyhole procedures. Why? The main problem is the size and design of the end-effectors, as they cannot be positioned too close to each other because of the risk of collision. This will be a major problem considering the small craniotomy in brain surgery, a nonissue in general surgery for example, where robotic arms are able to be positioned further apart. Another challenge may relate to the need for updated imaging and navigation considerations as neurosurgery is much more dependent on these systems for surgical planning and guidance, in comparison to general surgery. The da Vinci is not linked to these systems. Furthermore, the da Vinci system does not have a haptic interface, which would be considered much more important in brain surgery where the delicate nature of operation demands use of all available sensory input, that is, vision, sound, and touch of surgery.

6.8 Robotics for Microsurgical Assistance

In microsurgery, as a surgeon requires delicate and precise motions over a prolonged interval, a system, preferably robotic, that is able to stabilize the surgeon's upper extremities and reduce fatigue would rather be attractive and advantageous. Recently, an automatic adjustable surgical armrest, EXPERT was developed by the team that developed the NeuRobot and has shown promising results [112]. This system was designed such that the freely movable armrest follows the surgeon's arm and fixes in the adequate position automatically. In preclinical and clinical tests, it has shown to decrease surgeon fatigue as well as hand tremor and difficulty in performing neurosurgical procedures [112].

6.9 Robotics for Microsurgery

Although several telesurgical robotic systems with greater dexterity and enhanced tool manipulation abilities were designed and are being developed, few have reached clinical use for this purpose [121–125]. An example of such a technology that includes all these design considerations is the neuroArm [1, 4, 126], an MRI-compatible robot capable of both microsurgery and stereotaxy, designed and developed at the University of Calgary (Calgary, Alberta, Canada) in collaboration with MacDonald, Dettwiler and Associates (MDA, Canada) (Fig. 8). The collaborating company MDA designed and constructed the CanadArm II and Dextre for the National Aeronautics and Space Administration (NASA), and have translated some of the space technology into the neuroArm system's hardware and software capacities. The system includes a multisensory workstation and two remote slave-manipulators that can be mounted on a mobile base or on the operating table. The two arms have seven DOFs each and are designed to hold a variety

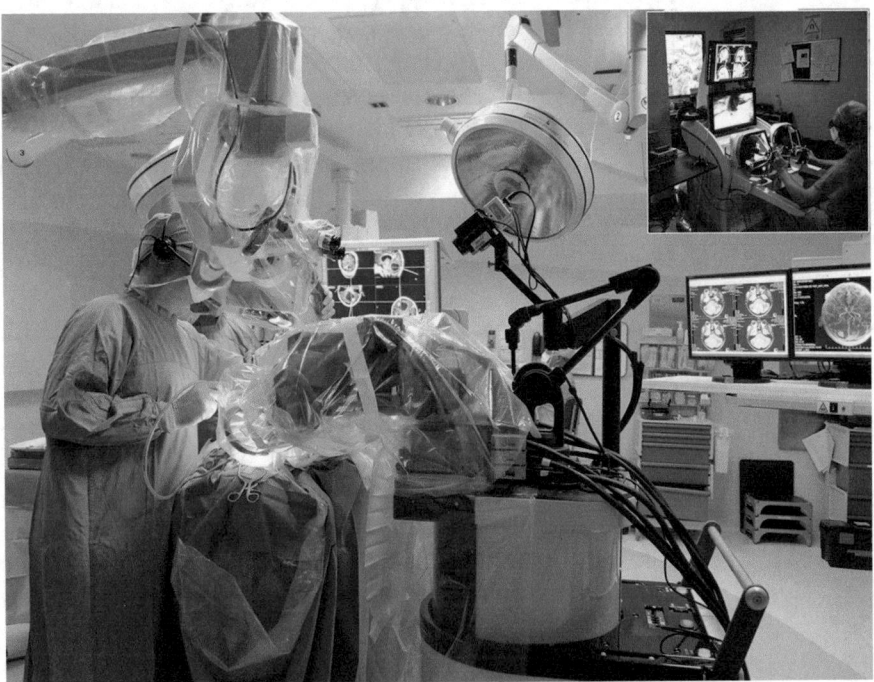

Fig. 8 The neuroArm telecapable robot developed for microsurgery (inset: the platform relocates the primary surgeon to a remote sensory-immersive workstation, here adjacent to the operating room)

of surgical tools including forceps, suction tubes, or laser systems [127]. For microsurgery, both manipulators are mounted on a mobile base in position, whereas for stereotaxy, the manipulator is positioned on a platform within the magnet bore registered to the magnet isocenter. Haptic feedback is provided in three translational DOFs from six-axis force/torque sensors that were specifically manufactured for neuroArm [128]. The sensory immersive workstation provides a responsive human machine interface that converges the 3D view of the operating site, the imaging data, a virtual display of the robotic manipulators in space, and specialized haptic hand controllers, for remote control of the manipulators in multiple DOFs [3]. The system has electronic tremor filters that results in smooth movement of manipulators regardless of surgeon's hand tremor. Moreover, neuroArm has motion scaling that allows the robot to move on a different scale than the surgeon's and force up-scaling that allows surgeon to feel very small forces, otherwise not possible for human hand to sense, or down-scaling large forces such that instability in the system is avoided. To address the potential risk of uncontrolled motion of the robotic manipulators at the surgical site, a foot pedal used as a *dead-man* switch and wired directly into the process logic controller, allows the surgeon to halt manipulators at any time in the event of unintended movement. In addition, neuroArm is equipped with a virtual simulator, which not only allows the practice of surgical tasks but also enable the surgeon to set electronic highways (no-go zones) or virtual fixtures, that can be preregistered for ensuring safety while approaching a target. Virtual fixtures have been shown to also increase surgical performance. After preclinical testing including rat model for microsurgery and cadaver model for the surgery within the bore of magnet [129], neuroArm has been successfully used for 80 clinical cases of microsurgery with varying pathology including brain tumor, abscess, radiation necrosis, and cavernous malformation [130–133].

7 Limitations

Despite the quantum leaps in technology, in particular software and more recently artificial intelligence, presenting an ideal robotic solution to the operating room, especially for microsurgery, remains a challenge. In a perfect world, one would well imagine a humanoid system, agile and intelligent, capable to assist or follow command with superb dexterity while also being able to make intelligent guess based on past memory such that complications are avoided ahead of time. This feat encompasses seamless and multifaceted integration of engineering principles that include mechanical hardware, software, computer science, electrical-control engineering, and machine learning. While robotic engineering, material science, and micromachining may have advanced, technical challenges of delay, backlash, inertia, stiffness, motion smoothness, and reproducing the human degrees of freedom in a robotic system do remain, despite advanced computation and software engineering. Why? Perhaps because what has been accomplished is only the beginning of understanding in the pathway toward creating the machine technology that works in tandem with surgeons in the operating room not only to assist or enhance the performance of surgery but to considerably alter the operating room workflow and enhance clinical outcome. Perhaps it is beyond capacity for physical-mechanical engineering to keep up with rapidly advancing software technology, computational power and machine learning. Another question is can robotics, similar to aviation industry, standardize surgery, such that care throughout specialties and centers is predictable, safe and standard? If yes, what are the pathways? As the world continues to pursue these goals to a better, robust and reliable robotic system, systematic, multifaceted, and stepwise approach to product development remains at the core, likely translating to increased cost or prolonged time for innovation.

8 Future Direction

Although neurosurgical robotics at its current stage, has opened doors for a wealth of scientific-engineering exploration, there has been a paucity in adopting the technology into everyday practice. While the theme of robotics appear attractive in the existing armamentarium of surgical tool technologies, several limitations remain with potential for further improvement. The move toward fully automated or semiautonomy system will continue, pulling away from simple execution of preprogrammed plans. CyberKnife already operates autonomously during its procedures, compensating instinctive motions of the patient such as those associated with breathing [29]. Some recent robotics for stereotaxy such as ROSA, NeuroMate, and SpineAssist also show similar levels of autonomy [19, 23, 37]. Raven II and PR2 robots, which are used for the execution of some confined tasks (e.g., laparoscopic suturing), and Smart Tissue Autonomous Robot (STAR) that incorporates 3D computer vision and sensors for robot navigation are among the tested platforms [134, 135]. Surgical procedure is not universal mainly because variances in anatomy and medical history prevail among patients, and therefore, numerous endeavors remain essential to account for the differences in order to introduce full autonomy. These could likely be addressed by the inclusion of artificial intelligence that has been attracting us for decades and has recently occupied a significant portion of advances and work by the information technology, corporations, and scientific community. It is no longer a distant unknown paradigm but is more a rapidly approaching realization, thanks to the leap forward in computer machine learning for memory, recognition, and intelligent interface that can think and operate like a human. The inclusion of artificial intelligence can also improve the technical and decision-making ability of surgeons by providing actionable real-time feedback to the surgeon through tactile, visual, and positional sensors to avoid adverse incidents and steepen the associated learning curve by automating the surgical evaluation process using console-feed videos [136]. The inclusion of speech recognition algorithms similar to Siri (Apple Inc., USA) can enable the surgeon to overrule the robot decision-making by giving the vocalized orders. Thanks to the video-enabled robotic procedures, image/video processing algorithms based on deep learning can arm the robots with autonomy in decision and operation, for example recognizing and navigating through nerves and vessels, and taking autonomous biopsies when desired. There is no need to teach the exact procedures to the robot, rather the abundance of data in terms of visual, sound, and tactile can make the machine learn the process from the examples, that is, unsupervised learning. Neurosurgery should be the pioneer in employing deep learning that constructs complex artificial neural networks similar to the human brain where the complex task (e.g., object recognition and tool navigation) splits up into simple subtasks across the whole network to get solved over time. This way an artificial brain can contribute to the treatment of a human brain. The machine learning capabilities of robots can bring them into a new world in which they are connected through 5G enabled Internet of Things (IoT) to communicate and learn from one another (Fig. 9). To fit this principle into future neurosurgical robots for high fidelity cognition and artificial intelligence toward semiautonomy of surgical procedures would require significant understanding of IT and software algorithms. In addition, incremental data input into the artificial neural network and computer processing capability for not only reproducing the ideal vision, sound, and touch of surgery but also guessing and decision-making based on past memory and experience of the robot. The challenges, however, remain in making the artificial intelligence to master the human surgeon considering the messy and unpredictable world of human characteristics and activities (e.g., fine and visually subtle maneuvers of surgeon or patient-specific tumor and organ appearances) [137].

8.1 End Effector and Instruments

Although the enhanced dexterity of a system such as neuroArm has the potential to conduct more complex surgical procedures, there are still several challenges to be addressed especially for microsurgical procedures that require accurate and precise motion. For example, several systems have been designed and tested for microvascular anastomosis, which requires the finest motion, few have been

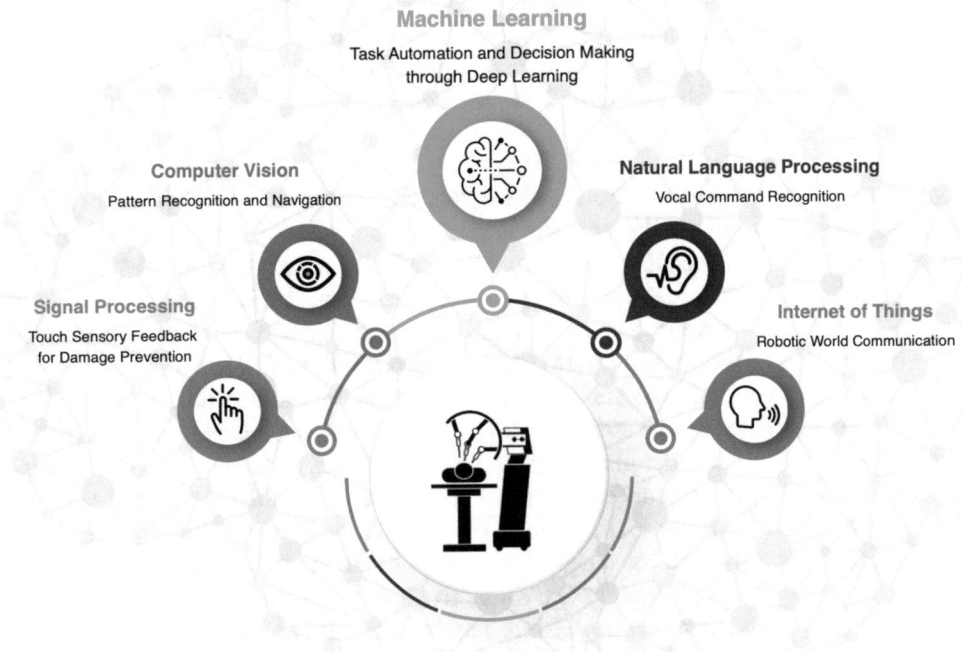

Fig. 9 The future direction of AI-incorporated robotic neurosurgery

superior to conventional freehand microsurgery [122–125, 138]. Improved end-effector for manipulators or tools capable of fine motion that could not be conducted by human hand, would be desirable. Moreover, by the nature of the central nervous system anatomy, the surgical corridor of neurosurgery is deep, narrow, and winding. Flexible, continuum robots or instruments which have redundant and increased DOFs are rapidly gaining attention due to their minimally invasive ability in reaching target locations through tortuous pathways maintaining fine motion [139–141]. The Flex Robotic System (Medrobotics Inc., Raynham, MA) [142], the i-Snake robot (Imperial College London, London, UK) [143], and the SPRINT (Single-Port laparoscopy bImaNual robot, EU consortium ARAKNES) [144] are good examples of such systems. In neurosurgery, the ROBOCAST/ACTIVE project (NearLab, Politecnico di Milano, Milan, Italy) utilized 13 DOFs for enhanced precision [19, 145], and the Raven project (Washington University, Seattle, USA) [146, 147] employ a flexible endoscope for improved visibility. Although of importance for increased DOF and its inverse relationship to the payload, it could be a promising feature for brain surgery as dissecting intracranial tissue such as brain, vessels, and arachnoid require gentle or low forces [148, 149]. Together with this trend, curved access to the deep brain lesions using a flexible steerable needle was also proposed [8]. Concentric tube robotic systems have also been applied for evacuation of intracerebral hemorrhage [150, 151] and endonasal skull base surgery [149, 152, 153].

8.2 Smaller Handheld System and Microrobot

Robotic systems and related instruments will inevitably become more lightweight and compact. Smaller systems able to provide accessible ability to deep lesion through narrow corridor without tissue injury on its trajectory, would likely come with reduced manufacturing cost. Moreover, smaller systems can be installed in confined spaces such as magnet bore or CT gantry [154]. Not only teleoperated systems but also shared-controlled systems that can work to support and augment surgeon's capability rather than to take the place of primary surgeon, are another attractive concept for microsurgery. One of the earliest examples of such a system was Steady Hand (Johns Hopkins University, Baltimore, Maryland), in which the surgical instrument is held by both the robot and operator, allowing for finer, tremor-free motion control of the instrument, while letting the surgeon define critical *no-go* areas to be avoided [155, 156]. In this context, untethered handheld devices may be favorable because they can be manipulated in exactly the same way as conventional surgical tools. To date, tremor suppression devices [157], force control system [158], and haptic feedback devices [159, 160] have been developed. Active guidance systems for microsurgery [161] and craniotomy [104, 105] are also being developed.

To reach the smallest section of the human anatomy such as subarachnoid space of the spine, further miniaturization of the system needs to progress. Like capsule endoscope, the microrobots technology is one of the newest research areas in surgical robotics [139, 141, 162]. Although there have been few studies that utilize microrobots, some investigators already reported promising results including intraocular or intracerebral drug delivery and arterial plaque removal [163]. Because of the small size, power transfer a challenge, for which external powering mechanisms are in development.

8.3 Advanced Imaging and Sensor Technology

Surgical robotics will continue to take advantage of *advanced intraoperative imaging*. This has pushed the merging of real-time imaging modalities with robotics since the mid-1990s. Due to the noninvasive, nonradioactive, and superior soft tissue contrast, MR compatible robotic systems have evolved. The NeuroBlate and neuroArm system are good examples of such systems. However, for intraoperative navigation, the combination of both preoperative and intraoperative imaging data and display is essential. Registration between different imaging modalities, scans, and times has been the focus of ongoing research [141, 164]. Although it remains challenging to track tissue deformation such as brain during surgery, for future systems, the technology of augmented reality can be included into robotics with overlay of pre- and intraoperative image data onto the exposed view of surgical field. Furthermore, cellular level of diagnosis and guidance could be incorporated, for maximum resection of malignant cells while preserving function. A long-held desire of neurosurgeons and for glioma is yet to be realized! For tumor margin interrogation, real-time in vivo identification and differentiation of infiltrated malignant cell from normal healthy brain tissue allows removing target cells more precisely. Photodynamic diagnosis using 5-aminolevulinic acid (5-ALA) and narrow band imaging are examples of such in vivo imaging [165, 166], notwithstanding some limitations such as local properties or low sensitivity. More recently, several novel technologies such as optical coherence tomography and confocal microendoscopy are being developed and have shown promising proof of concept [165, 167].

Superior *sensor* technology is highly wanted in the development of robotics. The neuroArm system signifies the importance of force sensing capacity that transmits the tool–tissue interaction back to the surgeon's hands during maneuvering the hand controller for tool actuation; however, few robotic systems have been successful in incorporating haptics. The sense of touch could be further augmented by using advanced miniaturized microprocessor sensors that will integrate the effect and provision of continuous data feed on to the workstation. Measurement of surgical forces and their relationship to tissue deformation will open a new area of research in basic science toward the understanding of tactile perception and its relationship to surgical decision-making. Design and development of miniaturized actuation systems are also crucial in invention of novel haptic systems to provide tactile and force feedback in micro scale for precise texture and skin pressure sensations.

8.4 Human–Machine Interface (HMI)

Together with better robotic design, suitable HMI is an essential prerequisite of robotic system for the synergistic and seamless control between surgeon and robots. Although majority of present teleoperated systems have utilized *off-the-shelf* hand controllers, a better HMI that is intuitive and user-friendly requiring minimal training would be necessary similar to the one being developed by Project neuroArm (Fig. 10). Especially in redundant robotic systems, the use of human hands is insufficient as it only allows for the control of only three DOFs at one time [139, 168, 169]. In comparison to conventional surgery, where surgeons benefit from multiple DOFs provided by fingers, wrist and forearm, most haptic hand controllers are designed to use a limited number of these DOFs to make it more intuitive and straight forward for operators

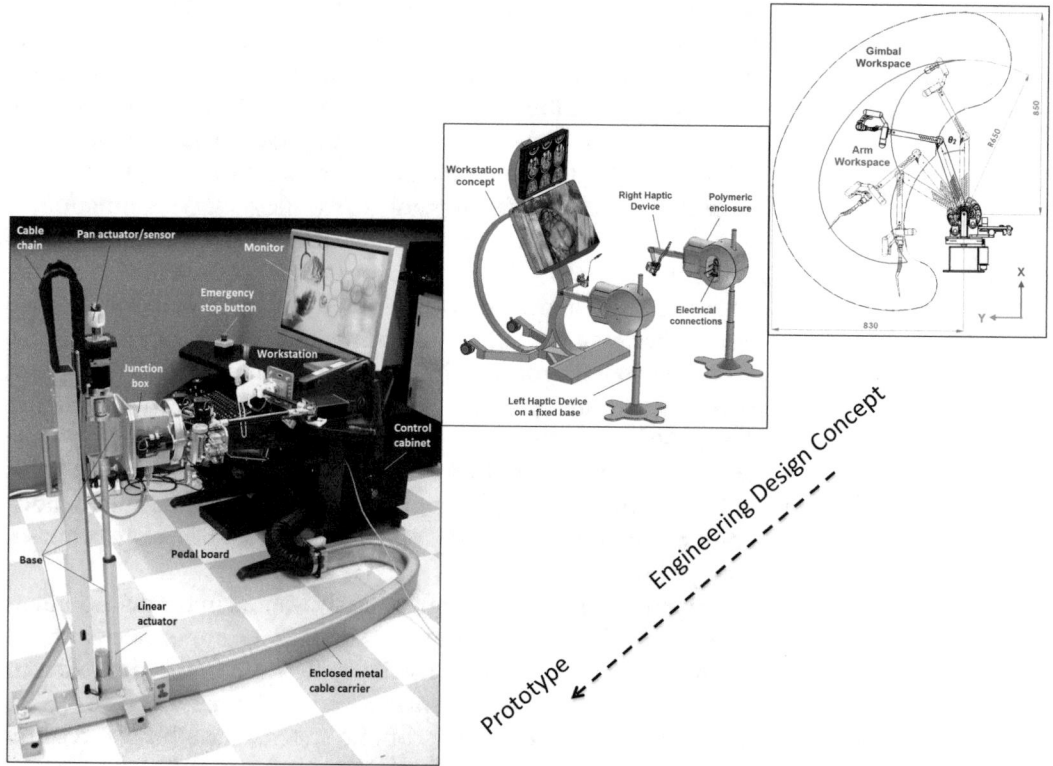

Fig. 10 Toward a unique microsurgery-specific haptic hand controller, from engineering concept to prototype and product development (US Patent WO/2019/126863 PCT/CA2018/000243)

to master. Further research is needed on design and development of advanced hand controllers with more dexterity and flexibility.

The tool–tissue interaction forces detected from the sensors installed on the robotic end-effectors need to be transmitted to the surgeon's hands correctly [3, 4]. Haptics or force feedback is provided in limited number of surgical systems such as neuroArm, and is still limited to kinesthetic forces provided in three dimensional Cartesian coordinate system. More research and development is required to provide additional DOFs of haptic feedback including torques or rotational forces. Considering and incorporating other elements of sense of touch such as temperature, smoothness or roughness, pressure, tickle, pain, vibration or even itch etc., would be of value for a surgeon at the workstation controlling a teleoperated surgical robot.

As the industry of high definition monitors is expanding exponentially, adapting to this advancement for improved vision from 3D camera, merging with external imaging data such as MRI would provide superior optics for robotics. Although cameras have an adequate quality and resolution today, more clarity and resolution might improve surgical performance specifically in robot-assisted paradigm.

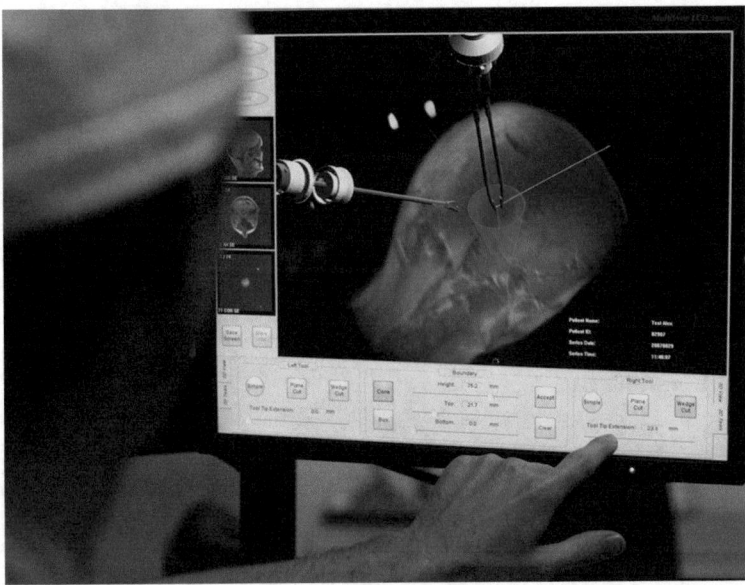

Fig. 11 The No-Go Zones/virtual fixture feature of the neuroArm system for increased safety

Not too dissimilar to any surgical technology, surgical robotics also has the risks of malfunction and failure. To circumvent these, numerous safety features must be considered early in development. No single point of failure must lead to loss of control of robotic system, that is, there remains minimal or no room for errors. *No-go zones or virtual fixtures* limit exerted force and speed, and the emergency stop systems that can halt robotic motion at any time of procedure needs to be incorporated into the system (Fig. 11) [3, 4, 132]. Use of redundant sensors and actuators is another solution for safety of the robotic systems. Incorporation of advanced control system design and schemes benefiting from robust control theories, stability analysis, passivity approach, and watchdogs are a few examples of possible solutions that can be considered. Because the use of robotic system requires additional training, an advanced simulation platform capable of case rehearsal, play forward, and playback would be also needed to enhance surgeon's training with shorter learning curves [4].

8.5 Cost

Although the future of robotic technology is promising and the possibilities that come with it exciting, one cannot ignore health care economy and high cost these systems presently incur [168, 170–172]. Similar to demand-and-supply philosophy, an increased demand and competitive manufacture of new robotic systems could potentially result in a considerable reduction of cost. However, increasing reports of adverse events to the FDA [172], reexamination of the cost-per-procedure, and the recent funding trend that surgical grant proposal are less likely funded compared to nonsurgical proposals [173–175] may negatively impact the development of new surgical robotic systems. Again, with the present-world scenario of interest and excitement of advanced technology robotics in medicine, the direction of surgical robotics will continue to expand. It is even conceivable that similar to initial abysmal cost of technology (e.g., NASA and how computers evolved for worldwide use), the spin-off results will continue to improve society, providing solutions to the fiscal health care dilemma, and accessibility to the masses.

9 Conclusion

Progress in neurosurgery has largely paralleled advances in technology, and this trend will continue to evolve. It is foreseeable that with contemporary technological advancement, neurosurgical robotics will become the hub of technology in the operating theater and it will be possible to interface imaging and surgical management with instantaneous global communication. Although neurosurgical robotics still needs considerable improvement and upgrade for global adoption, the future of this technology is promising and quite convincingly, inevitable.

Acknowledgments

This work was supported by Canada Foundation for Innovation, Western Economic Diversification and Alberta Advanced Education and Technology (Canada), and KANAE Foundation for the Promotion on Medical Science (Japan).
Conflicts of Interest Disclosure: Garnette R. Sutherland is listed on many of the founding patents of neuroArm.

References

1. Louw DF, Fielding T, McBeth PB, Gregoris D, Newhook P, Sutherland GR (2004) Surgical robotics: a review and neurosurgical prototype development. Neurosurgery 54(3):525–536; discussion 536–527

2. McBeth PB, Louw DF, Rizun PR, Sutherland GR (2004) Robotics in neurosurgery. Am J Surg 188(4A Suppl):68S–75S

3. Greer AD, Newhook PM, Sutherland GR (2008) Human-machine interface for robotic surgery and stereotaxy. IEEE/ASME Trans Mechatron 13(3):355–361

4. Sutherland GR, Wolfsberger S, Lama S, Zareinia K (2013) The evolution of neuroArm. Neurosurgery 72(Suppl 1):27–32

5. Mattei TA, Rodriguez AH, Sambhara D, Mendel E (2014) Current state-of-the-art and future perspectives of robotic technology in neurosurgery. Neurosurg Rev 37(3):357–366; discussion 366

6. Zamorano L, Li Q, Jain S, Kaur G (2004) Robotics in neurosurgery: state of the art and future technological challenges. Int J Med Robot 1(1):7–22

7. Herron DM, Marohn M (2008) A consensus document on robotic surgery. Surg Endosc 22(2):313–325; discussion 311–312

8. Engh JA, Minhas DS, Kondziolka D, Riviere CN (2010) Percutaneous intracerebral navigation by duty-cycled spinning of flexible

bevel-tipped needles. Neurosurgery 67 (4):1117–1122; discussion 1122–1113

9. Doulgeris JJ, Gonzalez-Blohm SA, Filis AK, Shea TM, Aghayev K, Vrionis FD (2015) Robotics in neurosurgery: evolution, current challenges, and compromises. Cancer Control 22(3):352–359

10. Nathoo N, Cavusoglu MC, Vogelbaum MA, Barnett GH (2005) In touch with robotics: neurosurgery for the future. Neurosurgery 56 (3):421–433; discussion 421–433

11. Marcus HJ, Seneci CA, Payne CJ, Nandi D, Darzi A, Yang GZ (2014) Robotics in keyhole transcranial endoscope-assisted microsurgery: a critical review of existing systems and proposed specifications for new robotic platforms. Neurosurgery 10(Suppl 1):84–95; discussion 95–86

12. Kwoh YS, Hou J, Jonckheere EA, Hayati S (1988) A robot with improved absolute positioning accuracy for CT guided stereotactic brain surgery. IEEE Trans Biomed Eng 35 (2):153–160

13. Drake JM, Joy M, Goldenberg A, Kreindler D (1991) Computer- and robot-assisted resection of thalamic astrocytomas in children. Neurosurgery 29(1):27–33

14. Benabid AL, Cinquin P, Lavalle S, Le Bas JF, Demongeot J, de Rougemont J (1987) Computer-driven robot for stereotactic surgery connected to CT scan and magnetic resonance imaging. Technological design and preliminary results. Appl Neurophysiol 50 (1–6):153–154

15. Varma TR, Eldridge PR, Forster A et al (2003) Use of the NeuroMate stereotactic robot in a frameless mode for movement disorder surgery. Stereotact Funct Neurosurg 80 (1–4):132–135

16. von Langsdorff D, Paquis P, Fontaine D (2015) In vivo measurement of the frame-based application accuracy of the Neuromate neurosurgical robot. J Neurosurg 122 (1):191–194

17. Varma TR, Eldridge P (2006) Use of the NeuroMate stereotactic robot in a frameless mode for functional neurosurgery. Int J Med Robot 2(2):107–113

18. Cardinale F, Cossu M, Castana L et al (2013) Stereoelectroencephalography: surgical methodology, safety, and stereotactic application accuracy in 500 procedures. Neurosurgery 72(3):353–366; discussion 366

19. De Momi E, Caborni C, Cardinale F et al (2013) Automatic trajectory planner for StereoElectroEncephaloGraphy procedures:

a retrospective study. IEEE Trans Biomed Eng 60(4):986–993

20. Abhinav K, Prakash S, Sandeman DR (2013) Use of robot-guided stereotactic placement of intracerebral electrodes for investigation of focal epilepsy: initial experience in the UK. Br J Neurosurg 27(5):704–705

21. Procaccini E, Dorfmuller G, Fohlen M, Bulteau C, Delalande O (2006) Surgical management of hypothalamic hamartomas with epilepsy: the stereoendoscopic approach. Neurosurgery 59(4 Suppl 2):ONS336–344; discussion ONS344–336

22. Barua NU, Lowis SP, Woolley M, O'Sullivan S, Harrison R, Gill SS (2013) Robot-guided convection-enhanced delivery of carboplatin for advanced brainstem glioma. Acta Neurochir 155(8):1459–1465

23. Haegelen C, Touzet G, Reyns N, Maurage CA, Ayachi M, Blond S (2010) Stereotactic robot-guided biopsies of brain stem lesions: experience with 15 cases. Neurochirurgie 56 (5):363–367

24. Deacon G, Harwood A, Holdback J et al (2010) The Pathfinder image-guided surgical robot. Proc Inst Mech Eng H 224 (5):691–713

25. Eljamel MS (2009) Robotic neurological surgery applications: accuracy and consistency or pure fantasy? Stereotact Funct Neurosurg 87 (2):88–93

26. Eljamel MS (2006) Robotic application in epilepsy surgery. Int J Med Robot 2 (3):233–237

27. Glauser D, Fankhauser H, Epitaux M, Hefti JL, Jaccottet A (1995) Neurosurgical robot Minerva: first results and current developments. J Image Guid Surg 1(5):266–272

28. Adler JR Jr, Chang SD, Murphy MJ, Doty J, Geis P, Hancock SL (1997) The Cyberknife: a frameless robotic system for radiosurgery. Stereotact Funct Neurosurg 69(1–4 Pt 2):124–128

29. Chang SD, Main W, Martin DP, Gibbs IC, Heilbrun MP (2003) An analysis of the accuracy of the CyberKnife: a robotic frameless stereotactic radiosurgical system. Neurosurgery 52(1):140–146; discussion 146–147

30. Yin FF, Zhu J, Yan H et al (2002) Dosimetric characteristics of Novalis shaped beam surgery unit. Med Phys 29(8):1729–1738

31. Lefranc M, Capel C, Pruvot-Occean AS et al (2015) Frameless robotic stereotactic biopsies: a consecutive series of 100 cases. J Neurosurg 122(2):342–352

32. Vadera S, Chan A, Lo T et al (2017) Frameless stereotactic robot-assisted subthalamic

nucleus deep brain stimulation: case report. World Neurosurg 97:762.e11–762.e14

33. Gonzalez-Martinez J, Bulacio J, Thompson S et al (2016) Technique, results, and complications related to robot-assisted stereoelectroencephalography. Neurosurgery 78 (2):169–180

34. Lefranc M, Le Gars D (2012) Robotic implantation of deep brain stimulation leads, assisted by intra-operative, flat-panel CT. Acta Neurochir 154(11):2069–2074

35. Calisto A, Dorfmuller G, Fohlen M, Bulteau C, Conti A, Delalande O (2014) Endoscopic disconnection of hypothalamic hamartomas: safety and feasibility of robot-assisted, thulium laser-based procedures. J Neurosurg Pediatr 14(6):563–572

36. Chan AY, Tran DK, Gill AS, Hsu FP, Vadera S (2016) Stereotactic robot-assisted MRI-guided laser thermal ablation of radiation necrosis in the posterior cranial fossa: technical note. Neurosurg Focus 41(4):E5

37. Gonzalez-Martinez J, Vadera S, Mullin J et al (2014) Robot-assisted stereotactic laser ablation in medically intractable epilepsy: operative technique. Neurosurgery 10(Suppl 2):167–172; discussion 172–163

38. Chenin L, Peltier J, Lefranc M (2016) Minimally invasive transforaminal lumbar interbody fusion with the ROSA(TM) Spine robot and intraoperative flat-panel CT guidance. Acta Neurochir 158(6):1125–1128

39. Lefranc M, Peltier J (2015) Accuracy of thoracolumbar transpedicular and vertebral body percutaneous screw placement: coupling the Rosa(R) Spine robot with intraoperative flat-panel CT guidance—a cadaver study. J Robot Surg 9(4):331–338

40. Lonjon N, Chan-Seng E, Costalat V, Bonnafoux B, Vassal M, Boetto J (2016) Robot-assisted spine surgery: feasibility study through a prospective case-matched analysis. Eur Spine J 25(3):947–955

41. Lefranc M, Peltier J (2016) Evaluation of the ROSA Spine robot for minimally invasive surgical procedures. Expert Rev Med Devices 13 (10):899–906

42. Roessler K, Ungersboeck K, Aichholzer M et al (1998) Image-guided neurosurgery comparing a pointer device system with a navigating microscope: a retrospective analysis of 208 cases. Minim Invasive Neurosurg 41 (2):53–57

43. Roessler K, Ungersboeck K, Aichholzer M et al (1998) Frameless stereotactic lesion contour-guided surgery using a computer-navigated microscope. Surg Neurol 49 (3):282–288; discussion 288–289

44. Willems PW, Noordmans HJ, Ramos LM et al (2003) Clinical evaluation of stereotactic brain biopsies with an MKM-mounted instrument holder. Acta Neurochir 145 (10):889–897; discussion 897

45. Kajiwara K, Nishizaki T, Ohmoto Y, Nomura S, Suzuki M (2003) Image-guided transsphenoidal surgery for pituitary lesions using Mehrkoordinaten manipulator (MKM) navigation system. Minim Invasive Neurosurg 46(2):78–81

46. Levesque MF, Parker F (1999) MKM-guided resection of diffuse brainstem neoplasms. Stereotact Funct Neurosurg 73(1–4):15–18

47. Pirotte B, Voordecker P, Joffroy F et al (2001) The Zeiss-MKM system for frameless image-guided approach in epidural motor cortex stimulation for central neuropathic pain. Neurosurg Focus 11(3):E3

48. Oppenlander ME, Chowdhry SA, Merkl B, Hattendorf GM, Nakaji P, Spetzler RF (2014) Robotic autopositioning of the operating microscope. Neurosurgery 10 (Suppl 2):214–219; discussion 219

49. Lollis SS, Roberts DW (2009) Robotic placement of a CNS ventricular reservoir for administration of chemotherapy. Br J Neurosurg 23(5):516–520

50. Bekelis K, Radwan TA, Desai A, Roberts DW (2012) Frameless robotically targeted stereotactic brain biopsy: feasibility, diagnostic yield, and safety. J Neurosurg 116(5):1002–1006

51. Benabid AL, Hoffmann D, Seigneuret E, Chabardes S (2006) Robotics in neurosurgery: which tools for what? Acta Neurochir Suppl 98:43–50

52. Eisner W, Burtscher J, Bale R et al (2002) Use of neuronavigation and electrophysiology in surgery of subcortically located lesions in the sensorimotor strip. J Neurol Neurosurg Psychiatry 72(3):378–381

53. Lollis SS, Roberts DW (2008) Robotic catheter ventriculostomy: feasibility, efficacy, and implications. J Neurosurg 108(2):269–274

54. Spire WJ, Jobst BC, Thadani VM, Williamson PD, Darcey TM, Roberts DW (2008) Robotic image-guided depth electrode implantation in the evaluation of medically intractable epilepsy. Neurosurg Focus 25(3): E19

55. Bertelsen A, Melo J, Sanchez E, Borro D (2013) A review of surgical robots for spinal interventions. Int J Med Robot 9 (4):407–422

56. Shweikeh F, Amadio JP, Arnell M et al (2014) Robotics and the spine: a review of current and ongoing applications. Neurosurg Focus 36(3):E10

57. Barzilay Y, Liebergall M, Fridlander A, Knoller N (2006) Miniature robotic guidance for spine surgery—introduction of a novel system and analysis of challenges encountered during the clinical development phase at two spine centres. Int J Med Robot 2(2):146–153

58. Dreval O, Rynkov I, Kasparova KA, Bruskin A, Aleksandrovskii V, Zil'Bernstein V (2014) Results of using spine assist mazor in surgical treatment of spine disorders. Interv Transpedicular Fixations 5(6):9–22

59. Hu X, Lieberman IH (2014) What is the learning curve for robotic-assisted pedicle screw placement in spine surgery? Clin Orthop Relat Res 472(6):1839–1844

60. Hu X, Ohnmeiss DD, Lieberman IH (2013) Robotic-assisted pedicle screw placement: lessons learned from the first 102 patients. Eur Spine J 22(3):661–666

61. Kuo KL, Su YF, Wu CH et al (2016) Assessing the intraoperative accuracy of pedicle screw placement by using a bone-mounted miniature robot system through secondary registration. PLoS One 11(4):e0153235

62. Macke JJ, Woo R, Varich L (2016) Accuracy of robot-assisted pedicle screw placement for adolescent idiopathic scoliosis in the pediatric population. J Robot Surg 10(2):145–150

63. Pechlivanis I, Kiriyanthan G, Engelhardt M et al (2009) Percutaneous placement of pedicle screws in the lumbar spine using a bone mounted miniature robotic system: first experiences and accuracy of screw placement. Spine (Phila Pa 1976) 34(4):392–398

64. Sukovich W, Brink-Danan S, Hardenbrook M (2006) Miniature robotic guidance for pedicle screw placement in posterior spinal fusion: early clinical experience with the SpineAssist. Int J Med Robot 2(2):114–122

65. Tsai TH, Wu DS, Su YF, Wu CH, Lin CL (2016) A retrospective study to validate an intraoperative robotic classification system for assessing the accuracy of Kirschner wire (K-wire) placements with postoperative computed tomography classification system for assessing the accuracy of pedicle screw placements. Medicine (Baltimore) 95(38):e4834

66. van Dijk JD, van den Ende RP, Stramigioli S, Kochling M, Hoss N (2015) Clinical pedicle screw accuracy and deviation from planning in robot-guided spine surgery: robot-guided pedicle screw accuracy. Spine (Phila Pa 1976) 40(17):E986–E991

67. Devito DP, Kaplan L, Dietl R et al (2010) Clinical acceptance and accuracy assessment of spinal implants guided with SpineAssist surgical robot: retrospective study. Spine (Phila Pa 1976) 35(24):2109–2115

68. Kantelhardt SR, Martinez R, Baerwinkel S, Burger R, Giese A, Rohde V (2011) Perioperative course and accuracy of screw positioning in conventional, open robotic-guided and percutaneous robotic-guided, pedicle screw placement. Eur Spine J 20(6):860–868

69. Keric N, Eum DJ, Afghanyar F et al (2016) Evaluation of surgical strategy of conventional vs. percutaneous robot-assisted spinal trans-pedicular instrumentation in spondylodiscitis. J Robot Surg 11(1):17–25

70. Onen MR, Simsek M, Naderi S (2014) Robotic spine surgery: a preliminary report. Turk Neurosurg 24(4):512–518

71. Schatlo B, Molliqaj G, Cuvinciuc V, Kotowski M, Schaller K, Tessitore E (2014) Safety and accuracy of robot-assisted versus fluoroscopy-guided pedicle screw insertion for degenerative diseases of the lumbar spine: a matched cohort comparison. J Neurosurg Spine 20(6):636–643

72. Hyun SJ, Kim KJ, Jahng TA, Kim HJ (2017) Minimally invasive, robotic-vs. open fluoroscopic-guided spinal instrumented fusions-a randomized, controlled trial. Spine (Phila Pa 1976) 42(6):353–358

73. Kim HJ, Jung WI, Chang BS, Lee CK, Kang KT, Yeom JS (2016) A prospective, randomized, controlled trial of robot-assisted vs freehand pedicle screw fixation in spine surgery. Int J Med Robot. https://doi.org/10.1002/rcs.1779

74. Kim HJ, Lee SH, Chang BS et al (2015) Monitoring the quality of robot-assisted pedicle screw fixation in the lumbar spine by using a cumulative summation test. Spine (Phila Pa 1976) 40(2):87–94

75. Liu H, Chen W, Wang Z, Lin J, Meng B, Yang H (2016) Comparison of the accuracy between robot-assisted and conventional freehand pedicle screw placement: a systematic review and meta-analysis. Int J Comput Assist Radiol Surg 11(12):2273–2281

76. Ringel F, Stuer C, Reinke A et al (2012) Accuracy of robot-assisted placement of lumbar and sacral pedicle screws: a prospective randomized comparison to conventional freehand screw implantation. Spine (Phila Pa 1976) 37(8):E496–E501

77. Roser F, Tatagiba M, Maier G (2013) Spinal robotics: current applications and future perspectives. Neurosurgery 72(Suppl 1):12–18

78. Schatlo B, Martinez R, Alaid A et al (2015) Unskilled unawareness and the learning curve in robotic spine surgery. Acta Neurochir 157 (10):1819–1823; discussion 1823

79. Schizas C, Thein E, Kwiatkowski B, Kulik G (2012) Pedicle screw insertion: robotic assistance versus conventional C-arm fluoroscopy. Acta Orthop Belg 78(2):240–245

80. Grimm F, Naros G, Gutenberg A, Keric N, Giese A, Gharabaghi A (2015) Blurring the boundaries between frame-based and frameless stereotaxy: feasibility study for brain biopsies performed with the use of a head-mounted robot. J Neurosurg 123 (3):737–742

81. Cleary K, Watson V, Lindisch D et al (2005) Precision placement of instruments for minimally invasive procedures using a "needle driver" robot. Int J Med Robot 1(2):40–47

82. Melzer A, Gutmann B, Remmele T et al (2008) INNOMOTION for percutaneous image-guided interventions: principles and evaluation of this MR- and CT-compatible robotic system. IEEE Eng Med Biol Mag 27 (3):66–73

83. Dorfer C, Minchev G, Czech T et al (2016) A novel miniature robotic device for frameless implantation of depth electrodes in refractory epilepsy. J Neurosurg 126:1622–1628

84. Minchev G, Kronreif G, Martinez-Moreno M et al (2016) A novel miniature robotic guidance device for stereotactic neurosurgical interventions: preliminary experience with the iSYS1 robot. J Neurosurg 126:985–996

85. Marescaux J, Leroy J, Rubino F et al (2002) Transcontinental robot-assisted remote telesurgery: feasibility and potential applications. Ann Surg 235(4):487–492

86. Mendez I, Hill R, Clarke D, Kolyvas G, Walling S (2005) Robotic long-distance telementoring in neurosurgery. Neurosurgery 56 (3):434–440; discussion 434–440

87. Meng C, Wang T, Chou W, Luan S, Zhang Y, Tian Z (2004) Remote surgery case: robot-assisted teleneurosurgery. Paper presented at Robotics and Automation, 2004. Proceedings. ICRA'04. 2004 IEEE International Conference

88. Liu J, Zhang Y, Wang T, Xing H, Tian Z (2004) Neuromaster: a robot system for neurosurgery. Paper presented at Robotics and Automation, 2004. Proceedings. ICRA'04. 2004 IEEE International conference

89. Cai M, Tianmiao W, Wusheng C, Yuru Z (2006) A neurosurgical robotic system under image-guidance. Paper presented at 2006 4th IEEE International Conference on Industrial informatics

90. Gao X (2011) The anatomy of teleneurosurgery in China. Int J Telemed Appl 2011:353405

91. Tian Z, Lu W, Wang T, Ma B, Zhao Q, Zhang G (2008) Application of a robotic telemanipulation system in stereotactic surgery. Stereotact Funct Neurosurg 86(1):54–61

92. Tian Z-M, Lu W-S, Zhao Q-J, Yu X, Qi S-B, Wang R (2008) From frame to framless stereotactic operation—cinical application of 2011 cases. In: Medical imaging and informatics. Springer, New York, pp 18–24

93. Wu Z, Zhao Q, Tian Z et al (2014) Efficacy and safety of a new robot-assisted stereotactic system for radiofrequency thermocoagulation in patients with temporal lobe epilepsy. Exp Ther Med 7(6):1728–1732

94. Mohammadi AM, Schroeder JL (2014) Laser interstitial thermal therapy in treatment of brain tumors—the NeuroBlate System. Expert Rev Med Devices 11(2):109–119

95. Sloan AE, Ahluwalia MS, Valerio-Pascua J et al (2013) Results of the NeuroBlate System first-in-humans phase I clinical trial for recurrent glioblastoma: clinical article. J Neurosurg 118(6):1202–1219

96. Hawasli AH, Bagade S, Shimony JS, Miller-Thomas M, Leuthardt EC (2013) Magnetic resonance imaging-guided focused laser interstitial thermal therapy for intracranial lesions: single-institution series. Neurosurgery 73 (6):1007–1017

97. Hawasli AH, Ray WZ, Murphy RK, Dacey RG Jr, Leuthardt EC (2012) Magnetic resonance imaging-guided focused laser interstitial thermal therapy for subinsular metastatic adenocarcinoma: technical case report. Neurosurgery 70(2 Suppl Operative):332–337; discussion 338

98. Brandmeir NJ, McInerney J, Zacharia BE (2016) The use of custom 3D printed stereotactic frames for laser interstitial thermal ablation: technical note. Neurosurg Focus 41(4): E3

99. Rahmathulla G, Recinos PF, Valerio JE, Chao S, Barnett GH (2012) Laser interstitial thermal therapy for focal cerebral radiation necrosis: a case report and literature review. Stereotact Funct Neurosurg 90(3):192–200

100. Habboub G, Sharma M, Barnett GH, Mohammadi AM (2017) A novel combination of two minimally invasive surgical techniques in the management of refractory radiation necrosis: technical note. J Clin Neurosci 35:117–121

101. Cunha-Cruz V, Follmann A, Popovic A et al (2010) Robot- and computer-assisted craniotomy (CRANIO): from active systems to synergistic man-machine interaction. Proc Inst Mech Eng H 224(3):441–452

102. Korb W, Engel D, Boesecke R et al (2003) Development and first patient trial of a surgical robot for complex trajectory milling. Comput Aided Surg 8(5):247–256

103. Eggers G, Wirtz C, Korb W et al (2005) Robot-assisted craniotomy. Minim Invasive Neurosurg 48(3):154–158

104. Follmann A, Korff A, Fuertjes T, Kunze SC, Schmieder K, Radermacher K (2012) A novel concept for smart trepanation. J Craniofac Surg 23(1):309–314

105. Kane G, Eggers G, Boesecke R et al (2009) System design of a hand-held mobile robot for craniotomy. Med Image Comput Comput Assist Interv 12(Pt 1):402–409

106. Zimmermann M, Krishnan R, Raabe A, Seifert V (2002) Robot-assisted navigated neuroendoscopy. Neurosurgery 51(6):1446–1451; discussion 1451–1442

107. Nimsky C, Rachinger J, Iro H, Fahlbusch R (2004) Adaptation of a hexapod-based robotic system for extended endoscope-assisted transsphenoidal skull base surgery. Minim Invasive Neurosurg 47(1):41–46

108. Zimmermann M, Krishnan R, Raabe A, Seifert V (2004) Robot-assisted navigated endoscopic ventriculostomy: implementation of a new technology and first clinical results. Acta Neurochir 146(7):697–704

109. Hongo K, Kobayashi S, Kakizawa Y et al (2002) NeuRobot: telecontrolled micromanipulator system for minimally invasive microneurosurgery-preliminary results. Neurosurgery 51(4):985–988; discussion 988

110. Goto T, Miyahara T, Toyoda K et al (2009) Telesurgery of microscopic micromanipulator system "NeuRobot" in neurosurgery: interhospital preliminary study. J Brain Dis 1:45–53

111. Takasuna H, Goto T, Kakizawa Y et al (2012) Use of a micromanipulator system (NeuRobot) in endoscopic neurosurgery. J Clin Neurosci 19(11):1553–1557

112. Goto T, Hongo K, Yako T et al (2013) The concept and feasibility of EXPERT: intelligent armrest using robotics technology. Neurosurgery 72(Suppl 1):39–42

113. Yang MS, Kim KN, Yoon DH, Pennant W, Ha Y (2011) Robot-assisted resection of paraspinal schwannoma. J Korean Med Sci 26(1):150–153

114. Perez-Cruet MJ, Welsh RJ, Hussain NS, Begun EM, Lin J, Park P (2012) Use of the da Vinci minimally invasive robotic system for resection of a complicated paraspinal schwannoma with thoracic extension: case report. Neurosurgery 71(1 Suppl Operative):209–214

115. Moskowitz RM, Young JL, Box GN, Pare LS, Clayman RV (2009) Retroperitoneal transdiaphragmatic robotic-assisted laparoscopic resection of a left thoracolumbar neurofibroma. JSLS 13(1):64–68

116. Lee JY, Lega B, Bhowmick D et al (2010) Da Vinci robot-assisted transoral odontoidectomy for basilar invagination. ORL J Otorhinolaryngol Relat Spec 72(2):91–95

117. Beutler WJ, Peppelman WC Jr, DiMarco LA (2013) The da Vinci robotic surgical assisted anterior lumbar interbody fusion: technical development and case report. Spine (Phila Pa 1976) 38(4):356–363

118. Lee JY, Bhowmick DA, Eun DD, Welch WC (2013) Minimally invasive, robot-assisted, anterior lumbar interbody fusion: a technical note. J Neurol Surg A Cent Eur Neurosurg 74(4):258–261

119. Marcus HJ, Hughes-Hallett A, Cundy TP, Yang GZ, Darzi A, Nandi D (2015) da Vinci robot-assisted keyhole neurosurgery: a cadaver study on feasibility and safety. Neurosurg Rev 38(2):367–371; discussion 371

120. Hong WC, Tsai JC, Chang SD, Sorger JM (2013) Robotic skull base surgery via supraorbital keyhole approach: a cadaveric study. Neurosurgery 72(Suppl 1):33–38

121. Arata J, Kenmotsu H, Takagi M et al (2013) Surgical bedside master console for neurosurgical robotic system. Int J Comput Assist Radiol Surg 8(1):75–86

122. Karamanoukian RL, Bui T, McConnell MP, Evans GR, Karamanoukian HL (2006) Transfer of training in robotic-assisted microvascular surgery. Ann Plast Surg 57(6):662–665

123. Le Roux PD, Das H, Esquenazi S, Kelly PJ (2001) Robot-assisted microsurgery: a feasibility study in the rat. Neurosurgery 48(3):584–589

124. Mitsuishi M, Morita A, Sugita N et al (2013) Master-slave robotic platform and its feasibility study for micro-neurosurgery. Int J Med Robot 9(2):180–189

125. Morita A, Sora S, Mitsuishi M et al (2005) Microsurgical robotic system for the deep surgical field: development of a prototype and feasibility studies in animal and cadaveric models. J Neurosurg 103(2):320–327

126. Sutherland GR, Latour I, Greer AD, Fielding T, Feil G, Newhook P (2008) An image-guided magnetic resonance-compatible surgical robot. Neurosurgery 62 (2):286–292; discussion 292–283

127. Motkoski JW, Yang FW, Lwu SH, Sutherland GR (2013) Toward robot-assisted neurosurgical lasers. IEEE Trans Biomed Eng 60 (4):892–898

128. Rizun P, Gunn D, Cox B, Sutherland G (2006) Mechatronic design of haptic forceps for robotic surgery. Int J Med Robot 2 (4):341–349

129. Pandya S, Motkoski JW, Serrano-Almeida C, Greer AD, Latour I, Sutherland GR (2009) Advancing neurosurgery with image-guided robotics. J Neurosurg 111(6):1141–1149

130. Maddahi Y, Gan LS, Zareinia K, Lama S, Sepehri N, Sutherland GR (2016) Quantifying workspace and forces of surgical dissection during robot-assisted neurosurgery. Int J Med Robot 12(3):528–537

131. Maddahi Y, Zareinia K, Gan LS, Sutherland C, Lama S, Sutherland GR (2016) Treatment of glioma using neuroArm surgical system. Biomed Res Int 2016:1

132. Sutherland GR, Lama S, Gan LS, Wolfsberger S, Zareinia K (2013) Merging machines with microsurgery: clinical experience with neuroArm. J Neurosurg 118 (3):521–529

133. Sutherland GR, Maddahi Y, Gan LS, Lama S, Zareinia K (2015) Robotics in the neurosurgical treatment of glioma. Surg Neurol Int 6 (Suppl 1):S1–S8

134. Schulman J, Gupta A, Venkatesan S, Tayson-Frederick M, Abbeel P (2013) A case study of trajectory transfer through non-rigid registration for a simplified suturing scenario. Paper presented at 2013 IEEE/RSJ International Conference on Intelligent robots and systems

135. Leonard S, Wu KL, Kim Y, Krieger A, Kim PC (2014) Smart tissue anastomosis robot (STAR): a vision-guided robotics system for laparoscopic suturing. IEEE Trans Biomed Eng 61(4):1305–1317

136. Baghdadi A, Hussein AA, Ahmed Y, Cavuoto LA, Guru KA (2019) A computer vision technique for automated assessment of surgical performance using surgeons' console-feed videos. Int J Comput Assist Radiol Surg 14 (4):697–707

137. Sayburn A (2017) Will the machines take over surgery? Bull R Coll Surg Engl 99(3):88–90

138. Saleh DB, Syed M, Kulendren D, Ramakrishnan V, Liverneaux PA (2015) Plastic and reconstructive robotic microsurgery—a review of current practices. Ann Chir Plast Esthet 60(4):305–312

139. Bergeles C, Yang GZ (2014) From passive tool holders to microsurgeons: safer, smaller, smarter surgical robots. IEEE Trans Biomed Eng 61(5):1565–1576

140. Burgner-Kahrs J, Rucker DC, Choset H (2015) Continuum robots for medical applications: a survey. IEEE Trans Robot 31 (6):1261–1280

141. Vitiello V, Lee SL, Cundy TP, Yang GZ (2013) Emerging robotic platforms for minimally invasive surgery. IEEE Rev Biomed Eng 6:111–126

142. Remacle M, Prasad VMN, Lawson G, Plisson L, Bachy V, Van der Vorst S (2015) Transoral robotic surgery (TORS) with the Medrobotics Flex System: first surgical application on humans. Eur Arch Otorhinolaryngol 272(6):1451–1455

143. Shang J, Noonan DP, Payne C et al (2011) An articulated universal joint based flexible access robot for minimally invasive surgery. Paper presented at Robotics and Automation (ICRA), 2011 IEEE International Conference

144. Piccigallo M, Scarfogliero U, Quaglia C et al (2010) Design of a novel bimanual robotic system for single-port laparoscopy. IEEE/ASME Trans Mechatron 15(6):871–878

145. De Momi E, Caborni C, Cardinale F et al (2014) Multi-trajectories automatic planner for StereoElectroEncephaloGraphy (SEEG). Int J Comput Assist Radiol Surg 9 (6):1087–1097

146. Hannaford B, Rosen J, Friedman DW et al (2013) Raven-II: an open platform for surgical robotics research. IEEE Trans Biomed Eng 60(4):954–959

147. Lum MJH, Friedman DCW, Sankaranarayanan G et al (2009) The RAVEN: design and validation of a telesurgery system. Int J Robot Res 28(9):1183–1197

148. Gan LS, Zareinia K, Lama S, Maddahi Y, Yang FW, Sutherland GR (2015) Quantification of forces during a neurosurgical procedure: a pilot study. World Neurosurg 84(2):537–548

149. Bekeny JR, Swaney PJ, Webster RJ III, Russell PT, Weaver KD (2013) Forces applied at the skull base during transnasal endoscopic transsphenoidal pituitary tumor excision. J Neurol Surg B Skull Base 74(6):337–341

150. Burgner J, Swaney PJ, Lathrop RA, Weaver KD, Webster RJ (2013) Debulking from within: a robotic steerable cannula for intracerebral hemorrhage evacuation. IEEE Trans Biomed Eng 60(9):2567–2575

151. Godage IS, Remirez AA, Wirz R, Weaver KD, Burgner-Kahrs J, Webster RJ (2015) Robotic intracerebral hemorrhage evacuation: an in-scanner approach with concentric tube robots. Paper presented at Intelligent Robots and Systems (IROS), 2015 IEEE/RSJ International Conference

152. Burgner J, Rucker DC, Gilbert HB et al (2014) A telerobotic system for transnasal surgery. IEEE/ASME Trans Mechatron 19 (3):996–1006

153. Swaney PJ, Gilbert HB, Webster RJ, Russell PT, Weaver KD (2015) Endonasal skull base tumor removal using concentric tube continuum robots: a phantom study. J Neurol Surg B Skull Base 76(02):145–149

154. Li G, Su H, Cole GA et al (2015) Robotic system for MRI-guided stereotactic neurosurgery. IEEE Trans Biomed Eng 62 (4):1077–1088

155. Taylor R, Jensen P, Whitcomb L et al (1999) A steady-hand robotic system for microsurgical augmentation. Int J Robot Res 18 (12):1201–1210

156. Fleming I, Balicki M, Koo J et al (2008) Cooperative robot assistant for retinal microsurgery. Med Image Comput Comput Assist Interv 11(Pt 2):543–550

157. Maclachlan RA, Becker BC, Tabares JC, Podnar GW, Lobes LA Jr, Riviere CN (2012) Micron: an actively stabilized handheld tool for microsurgery. IEEE Trans Robot 28 (1):195–212

158. Gilbertson MW, Anthony BW (2013) An ergonomic, instrumented ultrasound probe for 6-axis force/torque measurement. Conf Proc IEEE Eng Med Biol Soc 2013:140–143

159. Payne CJ, Yang GZ (2014) Hand-held medical robots. Ann Biomed Eng 42 (8):1594–1605

160. Yao HY, Hayward V, Ellis RE (2005) A tactile enhancement instrument for minimally invasive surgery. Comput Aided Surg 10 (4):233–239

161. Song C, Park DY, Gehlbach PL, Park SJ, Kang JU (2013) Fiber-optic OCT sensor guided "SMART" micro-forceps for microsurgery. Biomed Opt Express 4 (7):1045–1050

162. Nelson BJ, Kaliakatsos IK, Abbott JJ (2010) Microrobots for minimally invasive medicine. Annu Rev Biomed Eng 12:55–85

163. Ullrich F, Bergeles C, Pokki J et al (2013) Mobility experiments with microrobots for minimally invasive intraocular surgery. Invest Ophthalmol Vis Sci 54(4):2853–2863

164. Markelj P, Tomazevic D, Likar B, Pernus F (2012) A review of 3D/2D registration methods for image-guided interventions. Med Image Anal 16(3):642–661

165. Liu JJ, Droller MJ, Liao JC (2012) New optical imaging technologies for bladder cancer: considerations and perspectives. J Urol 188 (2):361–368

166. Stummer W, Pichlmeier U, Meinel T, Wiestler OD, Zanella F, Reulen HJ (2006) Fluorescence-guided surgery with 5-aminolevulinic acid for resection of malignant glioma: a randomised controlled multicentre phase III trial. Lancet Oncol 7 (5):392–401

167. Martirosyan NL, Cavalcanti DD, Eschbacher JM et al (2011) Use of in vivo near-infrared laser confocal endomicroscopy with indocyanine green to detect the boundary of infiltrative tumor. J Neurosurg 115(6):1131–1138

168. Marcus H, Nandi D, Darzi A, Yang GZ (2013) Surgical robotics through a keyhole: from today's translational barriers to tomorrow's "disappearing" robots. IEEE Trans Biomed Eng 60(3):674–681

169. Yang G-Z, Mylonas GP, Kwok K-W, Chung A (2008) Perceptual docking for robotic control. Paper presented at International Workshop on Medical imaging and virtual reality

170. Barbash GI, Glied SA (2010) New technology and health care costs—the case of robot-assisted surgery. N Engl J Med 363 (8):701–704

171. Faria C, Erlhagen W, Rito M, De Momi E, Ferrigno G, Bicho E (2015) Review of robotic technology for stereotactic neurosurgery. IEEE Rev Biomed Eng 8:125–137

172. Smith JA, Jivraj J, Wong R, Yang V (2016) 30 years of neurosurgical robots: review and trends for manipulators and associated navigational systems. Ann Biomed Eng 44 (4):836–846

173. Hu Y, Edwards BL, Brooks KD, Newhook TE, Slingluff CL Jr (2015) Recent trends in National Institutes of Health funding for surgery: 2003 to 2013. Am J Surg 209 (6):1083–1089

174. Keswani SG, Moles CM, Morowitz M et al (2017) The future of basic science in academic surgery: identifying barriers to success for surgeon-scientists. Ann Surg 265 (6):1053–1059

175. Rangel SJ, Efron B, Moss RL (2002) Recent trends in National Institutes of Health funding of surgical research. Ann Surg 236 (3):277–286; discussion 286–277

INDEX

Hani J. Marcus and Christopher J. Payne (eds.), *Neurosurgical Robotics*, Neuromethods, vol. 162,
https://doi.org/10.1007/978-1-0716-0993-4, © Springer Science+Business Media, LLC, part of Springer Nature 2021

Printed in the United States
by Baker & Taylor Publisher Services